UNDER FIRE AND UNDER WATER

THE JULIAN J. ROTHBAUM DISTINGUISHED LECTURE SERIES

UNDER FIRE AND UNDER WATER

Wildfire, Flooding, and the Fight for Climate Resilience in the American West

Bruce E. Cain

UNIVERSITY OF OKLAHOMA PRESS : NORMAN

Chapter 5 draws on research sponsored by the California 100 initiative, a joint project of the University of California and Stanford University.

Library of Congress Cataloging-in-Publication Data

Names: Cain, Bruce E., author.
Title: Under fire and under water : wildfire, flooding, and the fight for climate resilience in the American West / Bruce E. Cain.
Description: First edition. | Norman : University of Oklahoma Press, [2023] | Series: The Julian J. Rothbaum Distinguished Lecture Series ; volume 16 | Includes index. | Summary: "Explores the psychological, social, and institutional obstacles—beyond political partisanship—that have hindered decarbonization and slowed progress in meeting the challenges of extreme weather in the West"—Provided by publisher.
Identifiers: LCCN 2023012229 | ISBN 978-0-8061-9320-5 (hardcover : alk. paper)
Subjects: LCSH: Natural disasters—West (U.S.) | Environmental protection—Political aspects—West (U.S.) | Climatic extremes—West (U.S.) | Climatic changes—Economic aspects—West (U.S.) | Resilience (Ecology)—West (U.S.) | BISAC: POLITICAL SCIENCE / Public Policy / Environmental Policy | NATURE / Weather
Classification: LCC GB5010 .C35 2023 | DDC 363.34/720978—dc23/eng/20230816
LC record available at https://lccn.loc.gov/2023012229

Under Fire and Under Water: Wildfire, Flooding, and the Fight for Climate Resilience in the American West is Volume 16 in The Julian J. Rothbaum Distinguished Lecture Series.

The paper in this book meets the guidelines for permanence and durability of the Committee on Production Guidelines for Book Longevity of the Council on Library Resources, Inc. ∞

CONTENTS

FOREWORD

Among the many good things that have happened to me in my life, there is none in which I take more pride than the establishment of the Carl Albert Congressional Research and Studies Center at the University of Oklahoma and none in which I take more satisfaction than the center's presentation of the Julian J. Rothbaum Distinguished Lecture Series. The series is a perpetually endowed program of the University of Oklahoma, created in honor of Julian J. Rothbaum by his wife, Irene, and son, Joel Jankowsky.

Julian J. Rothbaum, my close friend since our childhood days in southeastern Oklahoma, has long been a leader in Oklahoma in civic affairs. He served as a regent of the University of Oklahoma for two terms and as a state regent for higher education. In 1974 he was awarded the university's highest honor, the Distinguished Service Citation, and in 1986 he was inducted into the Oklahoma Hall of Fame.

The Rothbaum Lecture Series is devoted to the themes of representative government, democracy and education, and citizen participation in public affairs, values to which Julian J. Rothbaum was committed throughout his life. His lifelong dedication to the University of Okla-homa, the state, and his country is a tribute to the ideals to which the Rothbaum Lecture Series is dedicated. The books in the series make an enduring contribution to an understanding of American democracy.

Carl B. Albert
Forty-Sixth Speaker of the
United States House of Representatives

UNDER FIRE AND UNDER WATER

A WICKED PROBLEM IN
A TROUBLED POLITICAL TIME

The early US explorers characterized the vast western territory they visited as a "Great American Desert," a region they deemed unsuited for much human habitation and economic development due to its arid climate and rugged topography.[1] The dim prospect of thriving in a land of such hardship deterred US expansion for several decades. But that changed over the course of the nineteenth and twentieth centuries as westerners found ingenious ways to make the area sustainable enough for large numbers of people to inhabit it over time. With considerable help from the federal government, they constructed dams, canals, pumps, and reservoirs to collect and store the region's water resources and to redistribute water from where it was plentiful to where it was not. They built railroad lines and later highways that linked the West to the East Coast. Technology had apparently tamed nature, turning a bleak, desolate land into a sustainable and inviting place to work and live.

That optimism about mastering nature's challenges has been badly shaken in the twenty-first century by the growing number and intensity of extreme weather events. Warming temperatures have triggered longer, more extensive droughts, straining natural and man-made water reserves and contributing to historically dangerous wildfires in the region's vast forested areas and rangelands. Sea level rise and stronger storms threaten to flood many low-lying Pacific coastal communities more extensively than in the past. The levees, dams, and seawalls that once protected and sustained westerners for decades now seem inadequate thanks to changing climate conditions. The American West must be reengineered for people and business to live more sustainably in it under the conditions of future global warming.

Adapting to harsher and more frequent weather changes is not just a technical challenge. It is also a formidable political problem, coming at a time when the US political system is itself stressed by rising polarization, globalization, social media, rising inequality, and resurgent racial tensions.[2] Climate policy has two principal components: first, lessening the anthropogenic contribution to climate change by decarbonizing the economy (i.e., mitigation), and second, protecting people and property to the extent possible from dangerous extreme weather events (i.e., climate resilience). Both tasks are inherently political. Science can produce technical solutions to slow global warming and protect against natural disasters, but politics and institutional incentives determine whether and when these projects will be implemented.

Because of the peculiar design of American government, technical solutions, no matter how clever or well designed, must still navigate a complex gauntlet of local, state, and federal political institutions.[3] Voters must be convinced to support and pay for what needs to be done. The political parties and interest groups must find solutions they can converge on. Agencies must act with a sense of urgency to enact the regulatory reviews that allow utility-scale clean energy facilities and protective infrastructure to be built in a timely way. It is a case not just of whether this all happens but also how quickly. If, as many scientists fear, there is a point of no return when the climate cannot be saved, greenhouse gas emissions must be limited as quickly as possible. And given the amount of past emissions, the United States also needs to build more protection from extreme weather to limit economic and human loss. In short, we need timely action on two fronts.

For a political system that was purposely designed to fracture power and divide responsibility for policy horizontally across branches of government and vertically across the levels of its federal system, urgency is a big ask. Partisan polarization has widened the gap between Democrats and Republicans about both whether and how to proceed on climate change policy.[4] And the political difficulties do not end there. Beyond the partisan divide are a host of collective action problems and behavioral constraints that slow progress on climate-related policies even when there is strong consensus that something must be done.

A MORE TROUBLED POLITICAL TIME

Politics also shaped nineteenth- and early-twentieth-century western development. Exploring, settling, and taming the American West were facilitated to a significant degree by government actions. Publicly funded water irrigation and transportation systems were required when private efforts did not succeed often or well enough. Inducing people to live in a harsher, more dangerous environment necessitated legislation that gave land away in exchange for taking on the burdens of settling and cultivating it. The US government needed to create agencies such as the Army Corps of Engineers with the technical expertise and capacity to build the levees and dams that could control periodic flooding and generate hydroelectric power.

But the political path to western sustainability today is considerably more complex and prone to veto by a larger number of sophisticated actors than it was in the nineteenth century. Without strong political will, policies that exact immediate costs for the sake of future benefits are inherently prone to delay and obstruction. Hence, utility-scale wind and solar facilities, the transmission lines that carry clean power to consumers and businesses, and the levees that protect coastal communities from rising seas can be stymied by opponents who know how to wring advantage out of America's arcane legislative, rulemaking, and permit processes.

Institutionalized obstruction is deeply embedded in America's political DNA due to the founding fathers' suspicion of concentrated power. This obstruction has found even fuller expression in US political institutions over time.[5] The system's many checks and balances and its strong federalism and ever-expanding number of elected offices have given rise to what Francis Fukuyama calls "vetocracy."[6] With an abundance of structural opportunities to obstruct major change and preserve the status quo, there is a strong inertial force in the American policy-making system. The bigger the change embodied in some policy proposal, the greater the inertial force that must be overcome; that is, the more policy steps that need to be taken, the more interest groups that must be co-opted or defeated and the more time for social media to turn public opinion against any ambitious departure from the status quo.

Emergencies of one sort or another sometimes create enough momentum to overcome the political system's natural inertia by opening policy windows of opportunity.[7] The prospect of foreign countries laying claim to western territories in the nineteenth century and the threat of being attacked on the Pacific front during World War II engendered a strong enough sense of public urgency to supercharge the task of settling and developing the American West.[8] When the western frontier was finally "closed" and World War II ended, the national commitment to western infrastructure withered over time.

The federal government is now perceived as the regulator rather than the enabler of infrastructure and growth. Heightened environmental awareness in the 1970s and small-government conservatism under Ronald Reagan in the 1980s ended the era of big water infrastructure and transportation projects in the American West, leaving it to the states and local governments to furnish new infrastructure.[9] As awareness of past environmental damage grew, the federal government and state governments enacted more legislation to clean up old messes and prevent new ones. The web of agencies involved in implementing and enforcing these laws became more complex over time.[10] Expectations about who should sit at the table when decisions are made also expanded, adding to the transaction costs and prospects for delay and obstruction.[11] The argument here is not that these were bad developments but only that the failure to set priorities over different environmental concerns means that we continue to relitigate the trade-offs between them with each new regulation and permit, creating a drag on climate policies and necessary infrastructure that the United States can ill afford.

One unsettled but increasingly critical question is how much emphasis to put on climate adaptation as opposed to mitigation. Initially, decarbonizing the electricity sector by switching from fossil fuels to clean energy in order to limit greenhouse gas emissions seemed like the only goal that mattered. While ample scientific evidence showed that global warming would eventually impact weather, the extreme weather manifestations were not yet dramatic enough during most of the twentieth century to push adaptation into the top tier of policy concerns. This has changed dramatically in recent decades. The incidence, expense, and ferocity of extreme weather events in the twenty-first century is increasingly hard to ignore.

Some in the environmental community initially worried that a stronger focus on resilience would divert resources away from the more essential decarbonization efforts.[12] But nature is forcing our hand with respect to building resilience. The failure to address extreme weather now will lead to a bigger liability bill that will need to be paid off by future generations. Emissions from large wildfires will offset the gains from reducing fossil fuel emissions. Extended droughts will lower the water levels in hydroelectric dams, lessening an important source of clean energy. And prolonged and widespread heat waves will strain the electricity grid beyond the capacity of existing energy-generating sources.

Adaptation, like mitigation, is affected by partisanship and ideology.[13] Whether or not people believe in the scientific evidence that the climate is changing due to human activities can shape their perceptions of extreme weather risk, as later chapters show. People who believe that global warming is not happening are more likely to discount unusually bad natural disaster events as simply random occurrences. In theory, observing an unusual number of extreme weather events over time should change minds but not if ideological screens and partisan loyalties get in the way. For too many people, the messenger has become more important than the content of the information. And trust in previously reliable sources of expertise such as the science community and academics in certain quarters has gone down. This lack of trust is reinforced by social media, which plays a heightened role in transmitting information and inducing attitudinal conformity among friends, family, and like-minded people.[14] Espousing a shared view signals solidarity with a group you identify with, while dissent risks social alienation.

Climate change skepticism, ideology, and partisanship explain a good deal of the difficulties of making climate and extreme weather policy. However, these factors do not explain the weak commitment and policy procrastination in blue communities (i.e., liberal Democrats). Why do those who purport to understand the consequences of climate change continue to build homes and businesses in areas prone to dangerous wildfires and floods? Why do environmental groups try to hold up the construction of utility-scale wind and solar that they know must be built up rapidly in order to replace fossil fuel generation? Why do people in upper middle–class progressive communities fight hard against building denser housing around public transit stops when they know that emissions associated with gasoline cars contribute significantly to global warming? Despite knowing that global warming would result in sea level rise, many liberal well-resourced coastal communities in California, Oregon, and Washington State are still talking the talk but not walking the walk on implementing protective measures. Why is this? Will it change as weather becomes even more extreme and the expenses of our neglect mount?

These are the central questions addressed in this book. A basic theme is that while climate change skepticism and political polarization presently play critical roles in climate-related policy debates, there are other psychological, social, and institutional obstacles to achieving both decarbonization and extreme weather resilience beyond partisanship. These include such factors as adhering to the belief that people have a right to live where they want regardless of the risks and social costs, NIMBYism,[15] complex permitting procedures, divergence over environmental priorities, a reluctance to pay for what is needed for solutions, and wide variations in community capacities. The convergence of some or all of these can block or slow progress on climate change resilience measures. My purpose is not to deprecate the real progress that has been made in some western states on reducing emissions and building resilience but instead to take a fuller look at the challenges ahead and to suggest some ways forward.

A WICKED PROBLEM IN TROUBLED TIMES

While the timing of our climate debate could not be worse due to rising partisan polarization in the modern era, limiting greenhouse gas emissions would have been a formidable political task in any case given the deeply carbonized state of modern life. Climate change policy, as some like to say, is inherently wicked.[16]

A policy's intrinsic characteristics can either ease or complicate the task of finding a resolution to climate change. Much depends on such factors as the policy's public salience, the feasibility of possible solutions, the interest groups aligned for or against it, and the linkage with other contemporary issues. Sometimes all these conditions are right for a

policy solution, but when they are not, this result is typically policy
delay or outright failure.

Consider a few examples of earlier environmental policy "successes."
One was the closing of the hole in the ozone layer. Ozone forms a protective layer in the stratosphere, blocking enough of the sun's ultraviolet rays
to prevent them from sterilizing Earth's surface and harming people.
Scientists in the mid-1970s discovered evidence of substantial damage
to the ozone layer over Antarctica caused by chlorofluorocarbon emissions associated with such consumer items as hairspray and shaving
cream products. Press coverage at the time highlighted the possible
negative health consequences of ozone depletion on people and animals,
generating enough public support in the United States for a Freon ban in
1979. Eventually that ban was extended internationally with the signing
of the Montreal Protocol in 1987. Four decades later the hole has been
closed, although unfortunately in a way that contributes to the current
greenhouse gas problem.[17]

Why was it politically possible to close the ozone hole? To begin with,
the threat was widely publicized and presented as an immediate physical
danger for humans if left unaddressed, not as a prediction of future global
warming that would change weather patterns. Also, there was little or
no partisan debate over the scientific claim that chlorofluorocarbons
were the cause of the ozone hole as there is today over the impact of
greenhouse gases on global warning. A clear scientific consensus emerged
that the chlorofluorocarbons in aerosol sprays, foams, and refrigerants
were the key factors.

More importantly perhaps, there was a relatively straightforward
solution—substituting hydrofluorocarbons for the chlorofluorocarbons—that would better protect the ozone layer. DuPont, the manu-
facturer of Freon, initially resisted the idea of a chlorofluorocarbon
ban but ended its opposition upon discovering the hydrofluorocarbons
substitute. From a political point of view, solving the ozone hole problem
adversely affected only a few industrial products, not the viability of large
economic sectors. It was only a success story, however, in a very specific
and qualified sense, since switching to hydrofluorocarbons ultimately
contributed to global warming.

A second illustrative environmental story is acid rain. Acid rain
is created when emissions from industrial sources such as coal-fired

power plants are transformed into acid particles that can be carried long distances in the wind before depositing on land and damaging forests, buildings, lakes, and the like. The environmental problems associated with acid rain were physically manifest and hence hard to deny. Scientists could measure high acidic levels in lakes and document the effects on forests and property with ample evidence. Despite considerable political pushback and a long policy gestation period, it was possible to make headway toward resolving the problem. Through upgrades in smokestack technology and novel regulatory mechanisms, the United States has been able to reduce its sulfur dioxide emissions by 93%, nitrous oxide emissions by 86%, and wet sulphate depositions by 70% since 1989.[18]

Acid rain was nonetheless a more formidable political challenge than closing the ozone hole was, for several reasons. The technical solutions were more expensive, and the industry resistance was stronger. Resolving the problem consequently took longer. Acid rain was first identified in 1872, and US scientists began tracking it in the late 1960s. This effort culminated in a 1980 law, the Acid Deposition Act, that established an eighteen-year monitoring and research program. When the Canadian government complained about cross-border pollution in 1982 during a presidential visit, President Reagan commissioned a panel to study acid rain and issue a report. Divisions in both the White House and Congress on this topic then prevented any further action until 1991, when Congress finally put in place a cap-and-trade system to control sulfur dioxide and nitrogen oxide emissions.

The acid rain problem foreshadowed many of the political problems associated with moving from fossil fuels to renewables to limit future greenhouse gas emissions. As an obvious rule of thumb, the more economic interests threatened by a given policy or regulation, the greater the political resistance. The acid rain problem threatened the economic prospects of the coal industry, which at the time was an even more critical component of the energy sector than it is today. Scaling from a few companies in the case of Freon to such a critical energy sector input as coal substantially increased the potential opposition and economic costs of remedying the problem.

Timing and context also mattered a great deal. The solution to the ozone hole was achieved at the end of the Carter administration, while the acid rain curb was largely fought out during the Reagan and

George H. W. Bush presidencies, when partisanship levels had begun to accelerate. President Reagan was a stronger conservative ally in environmental matters than Richard Nixon and Gerald Ford had been, reflecting the business community's growing concerns over the Clean Air Act and the Endangered Species Act.

A core general consideration that sets the most difficult environmental problems apart from routine ones is the length of time between a problem's causes and the public's awareness of it. If a train derails and spills well-known toxic chemicals into a river or if an oil well off the coast is not properly sealed and fouls the ocean, the consequences are immediately apparent. There is little or no disputing whether the problem exists. The acid rain issue required more time to acquire sufficient scientific evidence, communicate it to the public, and then generate enough political pressure to get a policy response.

The time delay aspect is even more severe with respect to greenhouse gas emissions. Carbon dioxide and other greenhouse gases had to accumulate at sufficiently high levels before the climate consequences kicked in at a politically salient level. It also took time to separate out the science behind the anthropogenic and natural contributions to extreme weather. Wet and dry cycles, for instance, have existed as far back as we can measure. Researchers needed time to disentangle natural cycles from human causes.[19] Warming per se did not seem as concerning to the public until scientists could connect it to rising sea levels and more severe natural disasters. Extreme weather effects also vary across areas and time, which further complicates making predictions. Communicating these predictions and winning the acceptance of a citizenry that had become heavily dependent on fossil fuels was destined to be a steep climb.

Proof of global warming for many people exists in their lived experience with rising temperatures, storms, droughts, floods, rising seas, and the like. Over a person's life span, weather changes can occur too gradually to raise concerns. Alternatively, even if some natural disaster is historically bad, it can be dismissed as an anomaly rather than the new normal. And how people experience a general warming trend is mediated by the baseline regional climate. People who live in colder climates or at higher elevations may not be bothered by warming trends. People who live far from rivers or rising seas may not experience flooding. Those who live on the East Coast rather than in the more arid West are less likely to experience dangerous wildfires and severe drought. Experiential variability leaves plenty of room for politically motivated skepticism and indifference.[20]

Nonetheless, these various personal experiences with extreme weather will be a critical factor in stirring adaptive actions. The indirect effects on taxes, insurance premiums, and energy costs may matter more in the long run. Extreme weather that is more frequent and hazardous will ultimately lead to rising costs and social upheaval that can no longer be ignored. This may eventually force even the most reluctant communities to become more resilient and more willing to limit their greenhouse gas emissions.

THE PLAN OF THE BOOK

Resilience challenges in the United States play out in different natural and social circumstances at the regional level. In some places, such as along the East Coast, extreme weather takes the form of more dangerous hurricanes with destructive winds and considerable flooding. In the western United States hurricanes are rare, but wildfires are more severe and droughts are more prolonged. In addition to differences in climate and topography, variations in the economy, demography, and institutions across regions factor into resilience politics as well. Chapter 1 considers the distinctiveness of the American West from the rest of the country in these dimensions.

I consider three distinctively western regional resilience challenges, namely drought, wildfires, and coastal flooding related to sea level rise and stronger Pacific storms. Each challenge highlights a particular theme about resilience policy. Chapter 2 considers the cyclic nature of droughts in the West and how the alternation between weather emergencies and normal conditions shapes the rhythms of drought policies. When conditions are bad enough, western state governments get their act together to impose cutbacks and water-saving efficiencies, but when the drought recedes from public view so does the impetus to prepare for the next one. Progress on preparing for droughts thus becomes episodic and incremental rather than continuous and always improving. Many western areas have made considerable progress in achieving greater water efficiency, but water use in the region is still much higher than in the rest of the country and likely to be further stressed by future global warming.

Chapter 3 examines the dangerous wildfires in the American West and what they reveal about risk perceptions in wildland areas. The threat of being burned out and possibly dying in a wildfire is traumatizing for those who live in or near wildland areas, but it is a concern for a relatively tiny fraction of voters as compared to those residing in western cities and suburbs. By comparison, the exposure to toxic smoke is much more widespread, but the harms to health are not widely understood. Consequently, local governments continue to develop properties in wildfire-prone areas, underappreciating the social costs of letting people live where they want to.

Chapter 4 turns to the solidly blue coastal communities that make a similar mistake, letting people reside in low-lying coastal areas that will likely flood more extensively due to rising sea levels. When forced by circumstances, wealthier cities have the capacity to protect their residents and businesses unilaterally but at the expense of poorer neighboring communities that must deal with the extra flooding caused by water displacement related to levees and seawalls.

Chapter 5 considers how successful adaptation and resilience will require cooperative measures across jurisdictional boundaries. Despite their divergent initial paths, water and energy systems are both converging toward a mix of localized and centrally coordinated governance to meet the challenges of adaptation and decarbonization. The book's conclusion offers some lessons from the experience of resilience policies to date. The appendix provides details of the surveys discussed in the chapters.

CHAPTER 1

DISTINCTIVE WESTERN FEATURES

Building resilience to future extreme weather events is a national challenge, but as just discussed, it will play out differently in various parts of the country. The region west of the 100th meridian is distinctive from the rest of the country in both natural and socioeconomic ways that impact resilience politics. Four features are particularly important: aridity, the abundance of both fossil and renewable energy resources, strong center-periphery tensions, and dispersed urban density.

The West had to be socially and physically engineered to become a heavily populated and economically thriving area. The prospect of longer droughts, more extensive wildfires, extreme heat, sea level rise, and flooding have undermined some of the older assumptions that made the West a sustainable environment in the twentieth century. But is the American West region up to the task? Can it function cooperatively given the diversity of its natural resources and politics?

Over the years many boundaries have been suggested between the eastern and western United States, reflecting either subjective cultural perspectives or physical features. Saul Steinberg's 1976 *New Yorker* cartoon "A View of the World from 9th Avenue" brilliantly satirized the parochialism of New York City residents by implying that in their minds, the border of the Hudson River and New Jersey separated their world from a vast, empty western landscape.[1] More conventionally, people tend to draw the line at prominent geographic features such as the Mississippi River or where the Rocky Mountains begin to ascend.

But from a climate perspective, the 100th meridian (or perhaps now the 98th due to global warming moving the precipitation line east) makes the most sense.[2]

The 100th meridian dividing line originated with the Powell geographic expedition in 1869 and its subsequent 1878 *Report on the Arid Region of the United States*.[3] Observing that the 100th meridian separated areas that had average precipitation levels above and below fifty-one centimeters per year, the report characterized the West as an arid region. Aridity as measured by modern science is more complex than just precipitation. It also includes how much water is taken off the surface into the air by temperature, wind speed, solar radiation, and relative humidity. By any measure, the American West is not uniformly arid or even semiarid. Some subregions such as the Pacific Northwest have above average levels of precipitation, while other areas in the Southwest fall well below. Nonetheless, as a broad generalization, aridity is far more prevalent in the American West than in the East.

How does this play out with respect to climate change politics? Baseline climate conditions shape the resilience challenges that different regions must face and the steps that need to be taken to endure future harsher weather. Many parts of the American West have always been more prone to droughts, extreme heat, water shortages, and wildfires than the rest of the country. Not only is there less precipitation in the West, but there is also more seasonal and yearly variation. Whereas precipitation in the East occurs throughout the year, the rainy season in many parts of the American West ends in the spring and does not resume until the late fall. This means that there must be adequate water storage to compensate for the absence of precipitation in the dry months.

Mountains in the coastal and interior West provided a natural method of storage, holding precipitation during the winter at higher elevations and then releasing it to lower elevations during the warmer summer months. In addition, dams, reservoirs, pumps, and irrigation systems managed the flow to prevent flooding and to meet the fluctuating seasonal demands of people, industry, and agriculture. Unfortunately, putting in place large gray water infrastructure inflicted considerable ecological damage on the areas that were flooded for reservoirs and dams. Consequently, water storage has become more controversial as environmental sensitivity has grown over time. Building and operating dams is much more difficult now than it was in the immediate post–World War II period even as the demand for storage rises due to population growth and increasing water stress related to continued global warming.[4]

Warming temperatures threaten to disrupt some of the core assumptions behind western water systems, diminishing the snowpack and causing rain runoff to occur during the winter as opposed to spring. This upsets the normal seasonal rhythms for storing and releasing water. Climate change potentially may also increase year-to-year precipitation variability, creating more uncertainty about storage capacity and plans. And if the volume of precipitation during the rainy years also increases in the future, it will likely stress the water storage infrastructure capacity to a breaking point, as happened in California with the Oroville Dam spillway in 2017.[5]

Water has always been a contentious issue in the West due to its scarcity in many parts of the region. Water issues are highly contentious across state lines and between different sectors of the region. States in the Colorado basin have quarreled for decades over their water allocations and will have even more reason to do so as their populations increase and water reserves drop.[6] Oklahoma and Texas agreed to share surface water that weaves in and out of their common border but then fought all the way to the US Supreme Court over where the water could be withdrawn from.[7] When periodic droughts lead to water-use cutbacks, this exacerbates tensions throughout the West between agriculture, urban-suburban residents, and environmental groups. When surface water supplies are restricted by dry weather, this incentivizes groundwater depletion as farmers turn to aquifer supplies to compensate for the lack of surface water. All of these long-standing problems will become more contentious with warmer temperatures and longer, more extensive droughts.

Climate change also worsens wildfire problems in the West. The yearly alternation between wet and dry years means that wet years produce more vegetation that then becomes hazardous flammable material during the dry summers. Wildfires are endemic to the western climate and in some ways are beneficial to forest health. But they exact a high toll in terms of lives, property, and health, as we shall discuss in greater detail in chapter 3. Wildfires are the product of a people problem (i.e., wanting to live in or near nature) with a natural world problem (i.e., harsher and more frequent drought conditions). The continued expansion into wildland-urban interface areas increases the odds of human-generated ignitions and extensive damage to life and property.

Given all these western climate challenges, one might ask why people choose to live in the West and whether they are aware of the risks of living there. To answer this, we turn to data from the 2019 western region survey conducted by the Bill Lane Center for the American West that asked residents about their experiences with and concerns about extreme weather conditions in their states. To test the uniformity of responses in the West, we divide the western region into five subregions: West Coast (Oregon, Washington, and California), Southwest/border states (Arizona, New Mexico, and Texas), interior states (Utah, Wyoming, Montana, Colorado, Nevada, and Idaho), Pacific states (Hawaii and Alaska), and plains states (North Dakota, South Dakota, Nebraska, Kansas, and Oklahoma).

When our western respondents were asked what they liked the most and least about their states, their responses could be organized into word clouds that represent the frequency of a given topic by the relative size of the word in the cloud.[8] Apparently, weather is what many westerners like the most about living in western states and what many others like the least. Heat dominates what people do not like even as many people move into the western or southern regions to escape the cold and snow in the northern and eastern portions of the country. We can learn a little more about this seeming contradiction by taking two states at opposite ends of the regional spectrum: California and Oklahoma. Californians see their weather as a plus, whereas Oklahomans view high temperatures as a minus. In other words, there is subregional divergence between the milder coastal and warmer interior western states' weather perceptions. Even within California, there are similar variations between Bay Area residents and those living in the Central Valley and desert areas of the state.

It is noteworthy that water scarcity does not seem to register as a negative feature of living in the West even though many states such as California and Oklahoma had considerable experience coping with severe droughts and water shortages in the second decade of the twenty-first century. The invisibility of water issues when there is no immediate drought crisis attests to the short public memory of drought experiences, a topic covered in chapter 2.

In any given year, there will be variation within the western region as to which states are experiencing a severe drought. From a political point of view, it would be easier to achieve regional solidarity on drought policy if the entire region experienced the same dry conditions at the identical time, but as it stands, subregional variation creates different windows of problem salience throughout the West. Subregional variation would matter less if these drought memories lasted longer, but time decays the general public's attention and interest when the rain returns. This pattern is particularly pronounced in the case of drought, as it rarely causes the level of death and property destruction in the United States that hurricanes and wildfires cause. Thus, the windows of opportunity for mobilizing public support for preventive drought measures open and shut too quickly for the political system to take the most forward-thinking and effective protective actions.

In the summer of 2019 when the survey was conducted, 15% of the regional respondents claimed that they had personally experienced severe drought conditions in the previous twelve months. As one might expect, there was considerable state variation across the western region in our numbers. Only 6% of those in Washington and 8% in Oregon, two states with historically wetter climates, said they had personally experienced severe drought in the previous twelve months, whereas the percentages were higher in the more arid and semiarid states such as Arizona (27%), New Mexico (15%), Texas (14%), and California (19%).

But while those who claim to have personally experienced severe drought in any given year are far from a majority, there is nonetheless a generally high concern throughout the western region about future droughts. When specifically asked about extreme drought in the next five years, 53% of westerners had either a great deal or moderate level of concern. This view was shared by majorities for all western subregions except Hawaii and Alaska.

There are similar patterns for expectations over the next five years about the related problems of extreme heat and wildfires. The levels of concern about both are above 50% throughout the region, with some modest subregional differences. Predictably, border state residents

(Arizona, New Mexico, and Texas) were most concerned about extreme heat, while wildfire concerns were most elevated in California, Oregon, and Washington due most likely to very bad fire seasons in the period 2017–19.

By comparison, westerners in general worry less about blizzards (25%), hurricanes (21%), and extreme cold (31%) even though all occur to some degree in specific areas due to snowstorms in mountainous areas and tropical storms along the Gulf Coast. Other natural disaster concerns are also primarily subregional, such as tornadoes (67% in the Plains), earthquakes (64% on the West Coast and 56% in Hawaii and Alaska), and severe flooding (56% in Plains states and 57% in border states).

In sum, the baseline climate of various states and subregions shapes their extreme weather challenge. Due to the baseline climate in the West, droughts, heat waves, and wildfires are more prevalent in the western region and will shape the resilience challenges there for years to come if the scientific predictions about climate change are realized.

DISTINCTIVE WESTERN FEATURES: DIVERSE ENERGY RESOURCES

Another distinctive feature of the American West is its energy resource mix. Here, the differences between East and West are more in degree than in kind. There are fossil and renewable energy resources in the eastern half of America, but they are not as abundant or diverse as they are west of the 100th meridian. Seven of the top oil reserve states and five out of the eight top natural gas reserve states either straddle or lie to the west of that demarcation.[9]

Western states also contain more potential for developing renewable energy than does the East. The cosmic compensation for aridity is abundant photovoltaic solar resources, especially in the Southwest. The West also has excellent wind resources in the Plains and interior states ranging from Texas up through Wyoming and Montana. US geothermal resources are abundant between the Rocky Mountains and the Pacific Ocean. From a political standpoint, having an ample supply of both fossil and renewable energy is a mixed situation, a blessing for potential economic value and a curse because it creates regional political tensions between extractive and green energy sectors.

All western states have significant potential for some form of renewable energy. Texas, Oklahoma, Wyoming, and California have both significant fossil and renewable energy resources. A few others such as Oregon and Washington have renewable resources (e.g., hydro and geothermal) but little in the way of fossil fuel deposits. No western states are as strongly skewed toward fossil fuels as are the eastern states of Pennsylvania, West Virginia, and Ohio. Ironically, despite having abundant green energy resources, eleven out of the seventeen western states joined the suit against President Barack Obama's Clean Power Plan. This opposition to clean energy attests to the economic power of the fossil fuel industry and the dominance of the Republican Party in the interior West and Plains states.

It might seem at first that partisanship and the state energy profiles are more relevant to the decarbonization debate than to the resilience question, but as we will show in some detail in chapters 2–4 partisanship and economic tensions can carry over to the adaptation debates as well. Climate change skeptics, as will be demonstrated in later chapters, are also less likely to report concern about extreme weather and more inclined to resist state-funded steps to protect people and property from these risks.

By political profile alone, one might reasonably suppose that a red state such as Oklahoma, where 65% of the voters supported Donald Trump for president in 2016, would be more resistant to building out renewable power than a blue state such as California, where the Trump vote was 32%. And indeed, in the interior and Plains states voters in the aggregate lean more toward climate skepticism on a whole battery of climate change opinions. But with respect to building out wind resources, economic activity does not strictly follow political leanings. Texas and Oklahoma are the two leading states in terms of wind energy installations, utilizing various state incentives such as new transmission lines, favorable tax breaks, and renewable portfolio standards to spur them on.[10]

Why? To some significant degree, it is because western red states with advantages in wind resources recognize the economic value of entering the renewable market. To put it another way, economic interest, when favorable, seems ultimately to trump political tribalism and ideology. To be sure, the mixture of both renewable and fossil fuel energy in any given western state also ignites political fights over the share of public subsidies and tax credits that go to different energy sectors. But the attribute of mixed energy resources differentiates western states from primarily gas and coal states such as Ohio, West Virginia, and Pennsylvania.

Furthermore, it is easier to find the land and permission to build out utility-scale wind and solar in interior West and Plains states than in the coastal blue areas such as California. For instance, it is nearly impossible to site utility-scale solar and wind in the affluent blue California counties on the western side of the state. This is partly due to the built-up residential areas, but even when there is open space, "not in my backyard" objections to the sight and sound of large wind turbines from highly educated pro–climate change policy Democrats forces most utility-scale wind projects out into the emptier rangeland and desert areas on the eastern side of the state. In these locales transmission lines must be built across the mountains and valleys, which increases the likelihood of wildfire ignitions from downed power lines. As discussed in chapter 3, wildfires cause enormous damage to property and endanger people's lives and health in several ways. Moreover, a bad wildfire year can offset a whole year's worth of greenhouse gas emission reductions achieved by transitioning from fossil fuels to renewables. In short, the resistance to building important large renewable projects with big land footprints is sometimes greater in the blue portions of the West than the red.

Of course, to be fair, the demand for renewables in the blue portions of the West largely drives the economic incentives to build in the red portions. The demand for renewable energy within the state borders and in the region creates entrepreneurial opportunities in many western red states to make money selling clean electricity to blue customers who want to use green energy. As the cost of wind and solar have dropped with technological advances, some consumers in red states have had to put aside their ideological aversions to green energy when they discover that they save money with rooftop solar by selling energy back to the grid and using less grid energy themselves. It is one thing to follow political values if doing so is profitable or at least costless, but it is another when there is no cost advantage for your politically symbolic preferences. Looking to the future, the West's underlying energy profile offers the possibility of trading across the western interconnected grid system to take advantage of its diverse energy portfolio to compensate for the fact that the sun does not always shine when the wind blows and vice versa. Complementarity across the grid system helps to solve the intermittency problem, providing more region-wide reliability to the electricity supply.

DISTINCTIVE WESTERN FEATURES:
A HEAVY FEDERAL PRESENCE

One of the more stunning but underappreciated facts about the American West is the amount of land owned by the federal government throughout the far and interior western states. This ranges from just below a third in Washington and Montana to over 50% in five states (Idaho, Oregon, Nevada, Alaska, and Utah). Residents and local officials who must deal with their Washington, D.C., landlord face a bewildering array of contact points and divergent government missions. Federal land in the West is managed by three separate departments (the Department of the Interior, the Department of Agriculture, and the Department of Defense) and four agencies (the Bureau of Land Management, the Fish and Wildlife Service, the National Park Service, and the US Forest Service) with diverse functional perspectives that range from making sure the resources are utilized for their economic value to wilderness preservation. In addition, the Bureau of Indian Affairs supervises federally recognized tribal reservations that have sovereign powers beyond the control of the states they are located in.[11] In short, the federal presence is large, varied, and complex.

How did this come to be? In large part, it is because the federal government played the central role in acquiring, settling, and developing western lands during the nineteenth century. Having amassed western lands through purchase, treaties, and wars, the federal government induced people to settle these frontier territories by offering opportunities to acquire land, building critical transportation systems such as railroads and later highways, providing armed security, and later constructing critical infrastructure such as dams, reservoirs, canals, highways, and bridges.

After the settlement period, the federal government retained many of these lands. This became a source of considerable controversy over time, especially with state residents who wanted to utilize the land for private development and commercial activities. Some westerners came to resent what they saw as increasingly burdensome environmental restrictions on federal lands. Political parties eventually sorted along pro- and antienvironmental lines in the late twentieth century, which led to abrupt shifts in federal land policy depending on which political party was in power in Washington, D.C. Local western resistance to federal control has taken various forms over the years: the Sagebrush Rebellion in the 1970s and 1980s over the privatization of federal lands, the Malheur occupation in 2016 over grazing fees, and, most recently, the alternating phases of expansion and reduction of the Bears Ears and Grand Staircase–Escalante National Monument lands under different presidential administrations.

While these issues receive a great deal of media attention, it is unclear whether these federal-state tensions are as strong as press reports seem to suggest. When we asked in our regional survey whether our respondents agreed or disagreed with the statement that the federal government owns too much land, only 28% of westerners agreed, 24% disagreed, and the rest did not know or had no opinion. Looking at individual states, the resentment of federal ownership was strongest in Alaska (56%), Utah (51%), and Nevada (40%). Concern about federal government regulations among westerners appears to be a bit higher (37% agreed vs. 28% disagreed).

The widest divergence on federal regulation and controls over western lands is across not subregions of the West but rather party lines in all western states. Forty-two percent of Republicans agree that the federal government owns too much land in their state versus 19% of Democrats and 32% of Independents. Similarly, 60% of Republicans think that the federal government has too many regulations affecting their state as opposed to 22% of Democrats and 41% of Independents.

Beyond these specific points of contention, the federal government has shaped adaptation politics in the West in other ways. Although the federal government constructed much of the infrastructure that made living in the West possible, it pulled back from this role during and since the Reagan administration. The Corps of Engineers and the Bureau of Reclamation still play an important role in maintaining and repairing floodwalls, dams, and other water infrastructure, but the era of big federally funded water and dam projects, for instance, is clearly over.

Despite some brief talk of bipartisan efforts at new infrastructure projects under the Trump presidency, nothing of great importance emerged until the Democrats came to power in 2021 and rammed through a bipartisan infrastructure bill with a minimal amount of Republican support. But even then, it will be up to the states to take the initiative to apply for and utilize these resources.

Another way the federal government shapes western policy is through the division of administrative responsibility across its various agencies. Major water and energy projects will often span the jurisdictional powers of several federal agencies. Policies across the federal agencies in terms of land use, freedom of movement by the public, and fire policy can vary substantially from each other as well as from state policies and private landownership. Seeking permits to move forward on projects on or adjacent to these lands often means separately consulting with several agencies over many months, raising the transaction costs and increasing the delays for water and energy infrastructure proposals, including those that might protect residents from extreme weather.

And finally, the way that the federal government incentivized the development of the West created an enduring legacy of private water rights and subsidies for scarce water. As a means of incentivizing land settlement and economic development, this proved to be successful but had the long-term effect of making it very hard to institute efficient water markets and equitable pricing systems. What happens in the past does not just stay in the past. The legacy of past policies, particularly when they are embodied as water rights, will shape the path of drought resilience in the American West well into the future.

DISTINCTIVE WESTERN FEATURES:
DISPERSED, HEAVY URBAN DENSITY

The Hollywood image of the American West is the frontier community, sparsely populated settlements separated by great distances. There is some residual truth to that iconic perspective. The Census Bureau classifies a county as "frontier" if its population falls below a given threshold and requires travel of a set distance by car and by foot to get to the nearest urban area.[12] While the number of designated frontier communities in the American West has decreased considerably since the start of the nineteenth century, the West still has far more frontier counties than the other US regions. Whereas six states in the eastern United States have no frontier counties according to the 2010 census, the American West contains five states in which 30% or more of their population live in frontier counties, including over half of Wyoming's and Montana's residents. While modern means of communication and transportation have lessened any lived sense of remoteness to a considerable degree in these areas, they still have distinct challenges in terms of health care (e.g., access to hospital and specialized doctors requires considerable travel), economic development (e.g., heavy dependence on extractive industries and tourism), and broadband access (e.g., companies see no profit in providing it).

The West still contains many open spaces, in part because it also has high urban density. Seven out of the ten states with the highest urban density are located there, led by California, (which might not surprise people) but also Nevada and Utah (which might). Twenty-one of the top twenty-five most dense urban areas in America are in California alone. Why is this so? Partly, it reflects the vast amount of state and federal public land in the American West set aside for parks, forests and ranchlands, and big commercial agricultural holdings that enjoy water subsidies. But topological features such as mountain ranges and coastal limits (e.g., Hawaii) also hem urban areas into confined spaces.

Dense urban areas are dispersed among sparsely populated and remote counties west of the 100th meridian, leading to very pronounced and abrupt changes in natural and political landscapes as one drives through western states. Eastern urban clusters are located nearer to one another than they are in the West. Democrats are concentrated heavily in these urban islands and Republicans are concentrated in the surrounding rural areas, while the suburbs serve as the pivotal purple electoral battlegrounds that ultimately decide national and state elections. This urban-rural split is manifested as an East-West division between coastal liberal and interior conservative voters in the Pacific states (i.e., California, Oregon, and Washington) but is also evident throughout Nevada, Texas, Arizona, Colorado, New Mexico, and Utah. Behind the political divide are such demographic forces as domestic white-collar migration, immigration, and higher birthrates in the nonwhite population.

These political trends have implications for extreme weather politics in the western region. Many climate challenges require regional and subregional solutions that will necessarily depend on cooperation among and with local communities and states. The adequacy of Colorado River water supplies for southern California and Baja California are potentially affected by upstream population growth in Arizona and Colorado. Wind resources in Wyoming could potentially compensate for the evening solar energy downturn in Pacific coastal states trying to lessen their dependence on gas. Wildfire smoke in the Pacific Northwest drifts east, carrying harmful particles in the air that pollute the interior West and endanger health. There is, in other words, inherent potential for both conflict and cooperation in the western region.

RESILIENCE IN THE WESTERN REGION

The initial climate change emphasis was on mitigation rather than adaptation. The uptick since 2000 in extended droughts, unusual heat waves, violent hurricanes, and wildfires has shifted the attention of both journalists and academics more toward problems of climate change resilience. Patterns are starting to emerge from the various extreme weather resilience efforts in western communities that could help us better understand why some efforts succeed while others fail. One obstacle is partisanship, which affects perceptions of extreme weather and shapes the politically feasible solutions that people are open to accepting. If voters fail to believe for ideological reasons that the climate is changing, this can diminish their overall willingness to approve of and pay for resilience measures. But the cumulative expense and damage of these events is mounting dramatically and may eventually become large enough to break down partisan filters and pave the way for more bipartisan solutions.

Several themes weave the four features discussed above into the narrative. The first is that the legacy of the federal government's role continues to shape resilience politics in the American West. Federal water subsidies and irrigation infrastructure projects that earlier fostered rapid agricultural development in the arid and semiarid West distort water prices and fuel bitter disputes between farmers, environmentalists, and residents over increasingly scarce water supplies. An overly zealous federal fire-suppression policy during the twentieth century created a dangerous accumulation of forest fuels that feed rapidly spreading superheated wildfires. Federal flood insurance encourages people to locate and rebuild in increasingly flood-prone areas. And many efforts at resilience require permission from federal agencies.

Another crosscutting theme is that the inherent physical characteristics of an extreme weather threat determine the width and length of a policy window. Westerners are aware that aridity is a common feature and know that future drought and wildfire events will likely be harsher. A policy window is an opportunity for policy action enabled by crisis. The more severe the crisis, the greater the public sympathy for the victims. This in turn neutralizes the inertia and resistance from interest groups and incentivizes politicians to expend resources to fund relief and reconstruction. But as media attention and the salience of a particular natural disaster wane, the window of opportunity starts to close, and the chances for significant measures diminish. In some cases, such as droughts and nuisance flooding related to gradual sea level rise, the impacts are too limited in scope or too episodic to mobilize sustained political action. In other cases such as wildfires, the extensiveness and the human displacement caused by an event open opportunities for more robust resilient countermeasures.

A third theme is that climate change resilience may require institutional as well as policy changes that span across urban and rural areas in the West. Many of the necessary actions spill across jurisdictional boundaries because the natural disasters themselves and the solutions require cooperation and collaboration between and across levels of government. Population growth and water use upstream affect the quantity and quality of water downstream. It does very limited good for one community to adopt fire-safe strategies such as vegetation management and stricter building codes if adjacent communities do not adopt such strategies. And seawalls that protect a given piece of the coastline can cause more erosion in other parts of the coastline. To overcome the tendency for communities to take care of their own at the expense of others, it is often valuable to create formal collaborations between local, state, and federal officials and to include outside stakeholders as well. In addition, policy problems sometimes require procedural and institutional innovations.

The next three chapters examine three specific types of extreme weather: drought, wildfires, and sea level rise. As mentioned earlier, each chapter highlights specific problems: weather cycles that create policy cycles (chapter 2 on droughts), dangerous events that emanate from risky decisions by people and local governments to expand in wildland areas (chapter 3 on wildfires), and resilience procrastination in liberal affluent communities (chapter 4 on sea level rise).

CHAPTER 2

CLIMATE CREEP AND
DROUGHT POLICY CYCLES

While natural disasters often stem from "bad" weather (i.e., storms), too many days of the "good" warm, dry weather that attracted people to live in the West over the past 150 years can cumulatively evolve into a drought emergency. Months of low precipitation during the wet months in late fall and winter can deplete water supplies to critical levels, eventually causing shortages and emergency water curtailments. Water deficits damage crops and increase political tensions between different economic sectors.

From a policy perspective, the cycle of drought management crises illustrates an important point about how steps toward resilience are sometimes made incrementally and not in one fell swoop. Water shortage emergencies open small windows of policy opportunity that lead to some water efficiency and storage measures. When the rains return and the water curtailments end, some of the water efficiencies (e.g., low-flow faucets and shower heads) are retained, but there is also backsliding toward old inefficient water habits. Without a visible ongoing water crisis, the will and momentum for long overdue fundamental changes in western water systems never develop.

Native Americans and then later European settlers learned how to cope with periodic droughts. But what worked in the past needs to adapt to the demands of future global warming and revisionist ecological concerns about big water infrastructure. Westerners need to develop new water supplies and forms of storage. They may also have to reconsider and revise legacy laws that prioritized agricultural needs over those of fast-growing urban sectors and environmental flows.

Western political leaders handle drought emergencies and water curtailments well for the most part but struggle to make much headway on a coherent water strategy, let alone one that can adequately handle future climate change conditions. Farmers continue to plant new orchards and developers continue to build new homes in water-stressed areas. Cities with resources invest in desalination and water recycling, while the poorest communities rely on bottled water and dwindling groundwater supplies. Public interest in fixing these assorted problems is neither stable nor strong enough to sustain policy momentum for very long. Consequently, water policy progress consists of gradually accumulated marginal improvements over time rather than a full embrace of what needs to be done structurally.

DROUGHTS IN THE AMERICAN WEST

While droughts have always been a recurring feature in the American West, they are now longer lasting and more frequent and have more devastating consequences in this region than in the rest of the United States. The National Integrated Drought Information System classifies droughts into five levels, D0 to D4, ranging from abnormally dry with minor effects on water supply and crops to exceptional droughts that result in water shortage emergencies and extensive crop and pasture losses. Droughts vary in their time, length, and area coverage. The longer a drought lasts and the wider the area affected, the greater the risk and harm of water shortages, vegetation damage, and uncontrollable wildfires.

Consider the stark contrast between a northeastern state such as Massachusetts and a western state such as California. The widely used Köppen climate classification consists of five main types (tropical, dry, temperate, continental, and polar) and further differentiates by precipitation and temperature. Massachusetts contains two different climates within its borders: a humid continental climate with warm summers and cold snowy winters in the state's interior and a humid subtropical climate with milder winters in the southeastern coastal areas. The state receives an average of forty-three inches of precipitation per year. Over the twenty-year period from 2000 to 2020, the longest drought episode in Massachusetts lasted forty-eight weeks in 2016–17, encompassing 52% of the state at a D3 level (i.e., extreme drought with major crop/pasture damage and widespread water shortages and restrictions).[1]

By comparison, California's climate profile is much more complex and geographically variable, as is typical in the American West. California has ten different Köppen climate types ranging from hot desert to tundra, but it is primarily characterized by four types of Mediterranean climate that all feature long summer seasons with little or no precipitation. Having ten distinct climate types is also not unusual for a western state. The Southwest (except for Texas), the Pacific West, and the interior West states all have ten or more distinct climates within their boundaries. The Plains states are less characteristically western in this sense, with fewer climate types than the rest of the region.

California's average yearly rainfall is less than half of Massachusetts's total (i.e., eighteen inches), and its experience with drought is much more extensive. Over the same twenty-year period, California's longest drought lasted 376 weeks (compared to 43 weeks for Massachusetts), with 58% of the state experiencing the highest drought level in 2014 (i.e., exceptional drought). Over two decades, Massachusetts experienced occasional short-lived and usually mild droughts, while California experienced significant droughts (D2 level and above) in at least 10% of the state, 50% over the entire twenty-year time span. People from Massachusetts moving to California would have had little or no prior experience with turning shower water on and off to save water while applying soap to their bodies, collecting bath water and rain to water their plants, or dealing with the odor of waterless urinals.

California likes to pride itself, as the writer Carey McWilliams famously put it, on being "the great exception."[2] But with respect to droughts in the West, California is less the exception than the rule. Every state located at or west of the 100th meridian experienced the highest official drought level (D4) at some point between 2000 and 2020. All but three (Washington, Alaska, and North Dakota) had at least one drought that lasted 250 weeks or more.

Living in the West has always meant coping with serious droughts, some of which were quite lengthy. The thirteenth-century Puebloan drought, for instance, lasted twenty-one years, leading to widespread crop failures, population displacement, and food rivalry among the

Native American tribes living in the interior West at the time.[3] Droughts and rainy seasons have also shaped popular perceptions of the US West. An 1820 army expedition led by Zebulon Pike and Stephen Long happened to coincide with a severe two-year western drought, leading the expedition's report to conclude that the American West was a "Great Desert" unsuitable for an agricultural economy. Unusually wet years in the 1860s and 1870s later fostered the opposite perception. The West, some thought, was a potential Garden of Eden that could become wetter with more farming because "rain follows the plow."[4] But as both cases illustrate, seasonal and annual variability have always shaped the distinctive challenge of living in the American West.

These weather patterns have policy and political consequences. The variability and seasonality of precipitation in the West puts a premium on storage: that is, water needs to be saved when it is plentiful in the winter months and released during the spring and summer months to compensate for the lack of rain. Moreover, water has to be transported from where it is plentiful and stored to where it needs to be used later. This requires large investments in infrastructure such as dams, reservoirs, canals, and pipes.

From a political point of view, authorizing, financing, and implementing large infrastructure is ostensibly popular across party lines, but it is much harder to achieve in the modern era than it was in the immediate post–World War II period. What people want and what they are willing to pay for are different matters. Moreover, the environmental consequences of large dams and reservoirs are better understood and appreciated now, and opponents are better organized and more adept at using environmental regulation to delay the construction and add to the expense of large water projects. This leaves incumbent politicians with the responsibility for authorizing the projects' costs while not reaping tangible political benefits until many years later if at all. In the past, the urgency of populating and defending the western territories provided the stimulus for breaking through these inertial forces. As military threats have receded, the inertial political forces have strengthened over time. Political leaders are more handicapped by modern politics at a time when they need to be most proactive in terms of building protection to new climate circumstances.

THE LEGACIES OF WATER POLICY

In order to populate the land that the United States acquired between 1783 and 1853 by treaty, conquest, and purchase, it was necessary to engineer the region's water system to enable large numbers of people to inhabit the area sustainably. This was both a technical and a policy challenge. The former required figuring out how to distribute and store water to compensate for the aridity that prevailed over much of the land west of the 100th meridian. The latter meant incentivizing citizens to take the risk of moving to and living and working in a more dangerous and inhospitable territory.

The main sustainability problem was not that the American West was one big desert lacking in water resources as some had supposed. It was instead making the West's climate and topographic variability work for them. Transporting water from where it was plentiful to where it was not necessitated an elaborate irrigation system of pumps, canals, and aqueducts. And due to the seasonality of western weather and the prospect of severe droughts, this also required a heavy public investment in dams and reservoirs to store large volumes of water.

Fortunately, the West's topography provided a key element for the region's water system. Winter precipitation deposited snow on the region's numerous mountain ranges. The cold winter temperatures at the high mountain elevations allowed the snow to accumulate into a winter water reserve that would release down the mountain with the spring and summer melt, providing water for human consumption and agriculture in the dry season. Later, the evolution of ever more powerful pumps made it possible to deliver surface water long distances and up over elevated lands and to draw groundwater from lower depths of the region's aquifers.

Seasonality and the prospect of drought necessitated the construction of dams, levees, and reservoirs to bank water and control river flows that regularly alternated greatly between wet and dry seasons. The population in the American West could not have grown as quickly as it did from the late nineteenth century on without the technical advances of the industrial revolution. Rain did not follow the plow, as some thought, but people followed the water infrastructure. Arizona and southern

California would not have blossomed into major population centers without modern water technology and distribution systems.

While this massive water infrastructure was a necessary condition for populating and developing the American West, it was not sufficient. The will to finance, build, and maintain water infrastructure had to come from either the private sector or the public sector or some combination of both. Initially, local communities and private companies provided the needed leadership and the capital. Some private efforts were quite successful. The Mormon community, for instance, was able to build a 5,000-acre water system around Salt Lake City. By 1910, so-called ditch companies and private corporations had constructed dams, canals, and water infrastructure covering 13.5 million acres across the western region.[5] But many private efforts failed, and grassroots advocacy to enlist the federal government's support grew in the late 1800s.

Federal government capacity in the first half of the nineteenth century was minimal. There was no Bureau of Reclamation, Department of the Interior, or Corps of Engineers as they exist today. The railroads, a powerful economic interest group at the time, supported land and water policies that enabled expansion and population growth because it was good for their business. Operating prior to modern lobbying, conflict of interest, and transparency laws, special interest influence was even stronger then than it is now. Eventually in 1902 as a result of this advocacy and aided by the growth of congressional representation from the West, President Teddy Roosevelt approved federal money for western irrigation and flood control. The federal government established what would eventually become the Bureau of Reclamation, the agency that was most responsible for funding and building today's extensive network of western dams, reservoirs, canals, and hydroelectric plans.

The legacies of this era are several and continue to shape contemporary drought management policy and politics. Above all, farmers gained privileged access to scarce water resources during this period, access they retain to this day. The share of water that goes to agricultural irrigation in western states, estimated to be 72% of all freshwater withdrawal in the seventeen western states, is considerably larger than in the rest of the country.[6]

This fact becomes more politically salient and controversial during droughts and water shortages, particularly when urban and suburban residents have to cut their personal water use to meet government-mandated cuts. In years when water is plentiful, agriculture's large water share is largely invisible to the public and contested primarily by nonprofit environmental groups. In the middle of a drought, however, when lawns turn brown or people have to curtail their outdoor and indoor water consumption, the press and the public are more inclined to take an interest in irrigation policy as it relates to water allocation. Does it make sense, people begin to ask, to subsidize water-intensive crops such as cotton and rice in arid and semiarid climate areas? Does agriculture's value merit the expense of supporting an extensive irrigation system when the farming contribution to gross domestic product in Arizona, New Mexico, and California is 5% or less?[7]

The fact that western irrigation commands so much water is the quintessential illustration of political path dependency, whereby actions during an earlier period of time lock in a supporting ecosystem of interest groups and political incentives that then sustain a particular policy or practice over many decades. The political pathway for western irrigation policy can be most simply described as follows. Agriculture and mining were the dominant western industries during the initial western settlement years. The nation's leadership had an expansionist mindset and sought to populate the West by ensuring that there would be sufficient water for both residential consumption and agricultural development. The federal government eventually stepped up to the plate to provide money and expertise. The Corps of Engineers (established in 1802) and the Bureau of Reclamation (established in 1902) built critical elements of modern western water infrastructure such as dams, reservoirs, levees, and canals up to the 1970s. Performing their missions assiduously, the agencies were rewarded with continued congressional budget support. Members of Congress in the mid-twentieth century, seeking to please their voters by bringing in the "pork," log-rolled big water infrastructure project appropriations without sufficient consideration as to the overall economic value and environmental costs entailed.[8]

As the expansionist and progrowth zeal waned and the environmental movement emerged in the second half of the twentieth century, the public became more aware of and concerned about the ecological damage associated large water storage projects. With the dual pressure of 1980s fiscal conservatism and growing environmentalism, the federal government pulled away from its leading role in funding and building major water infrastructure projects during the Reagan administration. But the legacy of the federal government's efforts left behind a deeply embedded policy commitment to western irrigation and storage that ultimately conflicted with the region's burgeoning urban-suburban water needs and emergent environmental concerns.

Even as western state economies have evolved away from agricultural dependence, the government continues to subsidize irrigation infrastructure and prices. The public underwriting of water infrastructure over time has proved to be larger than was initially expected. The US government thought it could recoup its capital expenses through water-user revenues but may have received 14% to 35% of those costs in the end, costs that were themselves underestimated at the time in order to ease the path of political approval.[9]

POLITICAL TENSIONS AND WATER ENTITLEMENTS

The federal government to this day also continues to sell irrigated water to farmers at well below market prices. While this may have made sense in earlier stages of western US economic development, critics worry about its distorting effect on water use, diverting a valuable natural resource from potentially higher-value uses. Public subsidies, they argue, undermine efficient water markets, lowering the incentive to use water efficiently and directing a scarce resource toward less economically valuable purposes.

A more widespread use of water markets could in theory reallocate water to more important uses, but the scale of such markets in the West at present is still quite limited. Moreover, the fact that profits from water trade markets accrue to a politically entitled group raises sensitive fairness issues among other water users. The stated public purpose of irrigation subsidies was to encourage agricultural activity, not to enable a subclass of businesses to enrich themselves by extracting rents on the basis of a prior entitlement that was established and is currently maintained through lobbying and political partisan connections.

An example of this problem can be found in California's Imperial Valley. Farmers there are able to sell subsidized water that they had purchased for $15 per acre-foot to San Diego City at $225 per acre-foot.[10] While such a trade allocates water to more valued uses, it also creates a large fairness gap between what farmers and consumers pay for the same water. The market might clear because the demand for water is so high that some urban entity is willing and able to pay for it at that price, but the perception that people and businesses are profiting from politically obtained advantages can become toxic when periods of severe water scarcity shine a bright media spotlight on such price disparities.

San Diego County, an area with a large military footprint and many vibrant, growing commercial areas, lacks adequate ground and surface water supply within its boundaries to provide for its residents and businesses even under current conditions, let alone future population growth or worsening drought conditions due to climate change. In 1991, the city got 95% of its water from contracts with the Los Angeles Metropolitan Water District. When the district in that year cut its state water project deliveries to the San Diego County Water Authority by 31% due to drought conditions, this served as a wake-up call to San Diegans. They resolved to diversify their water supply by making purchases of surplus irrigation water from the Imperial Valley and also developing locally treated water supplies.[11]

By 2018, the city had decreased its share of Los Angeles district water to 32% and increased its water purchases from the Imperial Irrigation District to 22%. San Diegans also took on the costs of desalination and stormwater recycling. They approved a plan to build a large desalination plant next to an existing power station. Completed in 2015, it currently serves about 400,000 county residents and supplies about 10% of the water supply. At the same time, San Diego increased its stormwater recycling to 5% of its supply. All of this makes sense in terms of building resilience to future drought conditions.

However, the costs for both desalination and recycling are high, over $2,000 per acre for desalination and $586 per acre-foot for stormwater recycling. As a consequence, the average monthly cost of water was $117 for a family four using 100 gallons per person per day (the San Diego average is 91 gallons per person) and rose to $198 per month for heavy water users (defined as 150 gallons per person and above). In short, prices increased substantially over the previous eight years as the city sought to diversify its portfolio. San Diegans pay considerably more for their water than residents in other western cities such as Salt Lake ($32), Fresno ($48),

Dallas ($51), and San Jose ($84). And when these costs are added to water utility prices, they act as a regressive tax that falls disproportionately on lower-income residents.

Desalination expenses in particular are hard to justify to ratepayers during years of normal or above-average precipitation. Santa Barbara, for example, decommissioned its desalination plant after operating it for three months in 1992, influenced by a rainy winter season and the availability of northern California water from a newly constructed aqueduct. But after experiencing water shortages during the 2013–17 drought, Santa Barbara then decided to renovate and reactivate its original desalination plant. As in San Diego, Santa Barbara residents chose higher costs in exchange for better water supply reliability in the face of rising climate variability.[12]

While allowing irrigation water to be sold by farmers to other users is a more efficient use of a scarce resource, market solutions have inevitable fairness consequences. Willingness to pay is partly a function of the ability to pay. Yearly increases of 6–7% per year are more easily absorbed by middle and upper middle–class communities than disadvantaged ones. And insofar as rising prices encourage conservation, this can lead to galling disparities in who conserves for whom.

During the 2013–17 drought, for instance, developers and wealthy individuals in Coachella Valley, an inland area in southern California that receives only three to four inches per year and where the average per capita water use is two or three times that of coastal areas, built six new major housing developments containing thirty thousand units, artificial lakes, and a private nineteen-hole golf course.[13] None of this is environmentally sustainable, but those with the means have the ability to procure scarce water resources regardless of the collective consequences. In short, when the new infrastructure projects related to desalinization, water recycling, and water storage are financed and controlled by local communities, the result is often a wider dispersion of water inequity and costs. Communities with the fiscal and governing capacity to apply for matching grants or recover costs through higher fees will find ways to deal with worsening drought conditions. Communities that lack those resources will do worse.

Another nineteenth-century western water legacy is the prevalence of appropriative water rights. The doctrine of prior appropriation severs the tie between land ownership and water rights as it exists in most eastern states and allows for water to be transported to nonriparian farmlands for irrigation purposes. This bestows a right to water based on those who initially claim it for a beneficial use. The doctrine developed during the earliest period of westward expansion when the federal government was offering land and other resources as an incentive to encourage population and economic development in the frontier lands.[14] The same progrowth logic underlies many of the so-called land acts of the period, such as the Preemption Act (1841), the Donation Land Act (1850), the Homestead Act (1862), the Timber Culture Act (1873), the Desert Land Act (1877), and the Timber and Stone Act (1878).

There are several key features of this doctrine. The first is the principle of first in time, first in use. The priority of water allocation is established by the seniority of the claim, which matters the most when water resources are stressed. Claims established earlier in time take precedence over later ones. To put it another way, there are different levels of water privilege. Those with junior rights get cut off during droughts before those with senior rights do. California Governor Jerry Brown, for instance, used his emergency powers in 2015 to suspend the irrigation water rights of those with claims dated to 1905 or later but permitted those with older rights to use or sell it as they saw fit.[15]

Second, the doctrine of beneficial use is largely defined in terms of economic value, which has become more problematic as environmental concerns have evolved since the second half of the twentieth century. Beneficial use means any beneficial use, not the most beneficial economic use or even the best public use considering all environmental, recreational, or aesthetic options. And to complicate matters further, there is a "use it or lose it" element to appropriative rights that encourages economically inefficient practices to maintain water rights.

Like subsidies, legal entitlements to irrigation water feed the growing political tensions between the agricultural, environmental, and urban-suburban consumers. Water rights are a matter of state law, and in all but three of the seventeen western states, legislative laws can be overridden by popular vote through direct initiative and referendum mechanisms. This means that a readily available way exists for majority opinion to abolish the current agricultural entitlements. Efforts to change water rights dramatically would of course likely incur legal challenges, but it is pretty clear that if drought and supply conditions become sufficiently dire, the system might be substantially altered.

There is some evidence that greater voter awareness of existing state water allocations leads to greater support for changing them. When California respondents were asked in the 2015 Hoover Institution–Bill Lane Center for the American West water attitude survey whether they supported reallocating water from agriculture to residential and business users, only 29% were in favor, while 39% opposed reallocating agricultural water. However, when a subset of them were first informed that agriculture gets 80% of the water used by humans, the numbers shifted to 47% in favor of reallocation and 30% opposed.[16]

Irrigation subsidies are also not popular. Forty-nine percent of the California respondents favored reducing or eliminating farm water subsidies versus 30% opposed. This issue divides California Republican voters even though agribusiness is closely associated with the Republican Party (40% in favor and 43% opposed), no doubt in part because public subsidies generally are unpopular with economic conservatives. As with many resilience matters, partisanship figures into public attitudes about farm subsidies and water shares but sometimes in nuanced ways.

THE CONSEQUENCES OF
WATER GOVERNANCE FRACTURE

A third legacy feature is the high level of fracture in water governance, which is rooted in the multiple attributes of water. As discussed earlier, water is for some purposes a private good that can be consumed, used, or traded by those with a right to it, subject often to some restrictions such as beneficial use. But water also has collective characteristics, including being shared sustainably by multiple users so as to prevent its depletion (i.e., a common pool resource); providing broad potential ecological, recreational, and aesthetic benefits (i.e., public goods); and leading to broadly shared negative consequences such as bad water quality and pollution (i.e., negative externalities). Governments perform corrective functions that correspond to these specific water attributes: they can authorize, supervise, and enforce water markets; incentivize or enforce restrictions to prevent the depletion of a common pool resource such as an aquifer; or legislate and regulate to prevent negative externalities that arise from human water uses such as damage to fish and wildlife.

Complicating matters further, these different water functions are divided and shared both within and across various government levels. When multiple agencies at the same level of government have jurisdiction over pieces of a policy or regulatory process, it is called horizontal fracture. Horizontal fracture has increased as the scope and powers of US government at all levels have expanded over time and as legislative implementation is parceled out to different federal and state executive agencies. Parallel laws in some states have proliferated the number of agencies also involved in regulating proposed local, state, and federal water projects. Expanded horizontal fracture at all levels of government means more consultations with various state and local actors, which in turn increases the time, effort, and resources necessary to initiate a water infrastructure project.

Vertical fracture is when regulatory authority is parceled across federal, state, and local authorities. In theory, states control inland waters. But this seemingly clear line of state sovereignty has blurred as a consequence of federal environmental laws such as the National Environmental Policy Act, the Clean Water Act, and the Endangered Species Act. Obama administration executive actions aimed to expand federal authority over inland water under the Clean Water Act. Subsequently, the Trump administration sought to reverse them. Whether the federal government should interfere in the states' "absolute right to all their navigable waters" is now a highly polarized issue in the American West.[17]

Until the 1970s, the federal government promoted western growth through the financing and construction of water infrastructure. But since then, the federal government has essentially traded its infrastructure leadership for an enhanced regulatory role in the American West. Resentment over water regulation now melds with similar Sagebrush Rebellion–like concerns over the size of national monuments, the payment of grazing fees, and federal ownership of so much public land in western states. These seething political tensions that increasingly fall along partisan lines prevent Congress and the president from mediating state water disputes. When western states have disagreements over how to share water that flows across boundaries, they have to sort things out by themselves through compacts or expensive litigation.

While water politics often play out along predictable partisan lines at the state and federal levels, it is usually less partisan at the local level either because the elected offices are nonpartisan or because the local politics are solidly red or blue. However, this does not mean that the issues are any less complex.

To put it simply, building water infrastructure has become more difficult as the thicket of relevant agencies and regulations has increased. This is not to derogate the purposes of either the laws or the agencies that populate this thicket but only to say that federalism plus the division of labor and specialization logic of modern bureaucracies have over time created a maze of decision points that must be navigated to initiate a new water project. Democrats and Republicans acknowledge that there must be easier paths to renewing America's infrastructure, but they have polarized ideas about how to achieve this. The middle road is to find ways to expedite permitting processes without eviscerating the laws themselves.

More generally, the regulatory thicket is also harder to negotiate because we have also expanded opportunities for public participation in administrative procedures for the purpose of eliciting more granular input about local environmental impacts.[18] The implicit democratic assumption behind enhancing public feedback is that there will be no selection bias in it, that is, that the individuals and organizations who show up to the hearings or provide written comments are a representative sample of their local communities. Alas, that is all too often not true. While there are situations where the impacts are potentially so widespread as to motivate the community, all too often the public input is skewed toward those with specific interests at stake (e.g., NIMBY groups) or who are well organized and better resourced.[19] The problem is not their participation per se but rather the absence of others who may not immediately recognize their interests in a project or lack the time and capacity to act effectively.

This is counterintuitive perhaps, but sometimes in democracies more is less. That is, providing too many public participation opportunities can overwhelm the capacity of many citizens to utilize them effectively.[20] The multiple stages of permitting and evaluation provide numerous opportunities for objection and obstruction. These are called veto points.[21] They can be useful when they lead to more careful consideration of a proposal's impacts and externalities but can tip into dysfunction when the goal is to obstruct a majority-preferred resolution or prevent closure on an issue altogether.

BREAKING THE LOGJAM

The legacies of past western water policies have created a considerable amount of institutional and political inertia that shape whether, how, and when current drought policy is made. The flaws and tensions of a water system built on agricultural preferences and fractured water governance are largely invisible to the public except during severe droughts. Political incentives favor decisive action during an emergency and indecisive action the rest of the time. Without intense media scrutiny or the discomfort of drought restrictions, the forces that keep the status quo in place are mostly unchallenged, and efforts toward structural change flounder.

Interest groups and powerful economic sectors operate most effectively in the darker corners of democracy, away from the media glare and public scrutiny. Without issue salience, it is politically difficult to overcome the inertial interests that keep problematic policies in place. No matter how often the public is admonished about the need for better drought resilience, the same structural problems persist: the lack of sufficient storage, the inability to plan rationally across jurisdictions, and festering tensions over water use between urban-suburban, agricultural, and environmental sectors.

Incremental progress is possible, such as achieving lower per capita water usage in the West over time. Understanding why and when progress is politically possible is the key to plotting a more political effective water policy for the American West. In the sections that follow, we will first look more closely at how political incentives shape water policies and whether drought crises create opportunities for major structural change.

POLITICAL INCENTIVES AND
DROUGHT PREPARATION

Drought management consists of three core elements that correspond to the period before, during, and after a severe drought: preparedness, crisis management, and recovery. Drought preparedness entails building the infrastructure capacity to store, reuse, conserve, and reallocate water in order to minimize crop damages and potable water deprivation during extended dry periods. Crisis management encompasses the steps taken during a water shortage emergency to conserve water and meet basic water needs. And recovery refers to financial relief for drought damage and efforts to rebuild the water supply. Recovery from the last drought bleeds into preparation for the next one, creating a recurring policy cycle.

While much is known about the relative technical merits of different drought-resilience measures, our focus is on their politics. Government policy in one phase of the drought cycle can determine what needs to be done in another. For instance, inadequate preparation can make drought crisis management more difficult and expensive. Steps taken at the height of the drought, such as better conservation and efficiency, can be incorporated permanently into the recovery and preparation strategies or abandoned when no longer immediately necessary. Squandering opportunities for improvement during the recovery phase can lead to higher drought relief expenses down the road.

In general, governments are best at handling the emergency and relief phases of natural disasters and weakest at taking the necessary steps to prepare for the next drought.[22] Unlike hurricanes and wildfires, droughts are slowly evolving events that only become manifest emergencies if the dry conditions persist long enough. Whereas the time frame for hurricanes and wildfire disasters is days or weeks, drought events develop and extend over months and years. Public officials have ample time to consult with experts and decide about issuing drought emergency proclamations. Frontline personnel in a drought do not have to navigate fast-moving, dangerous situations to save lives and protect property. Droughts in the American West can of course cause widespread economic loss and deprivation, but rarely do they result in widespread death and destruction. Because they are less hazardous, the perceived danger is less, and the political risks associated with failing to take appropriate preventive actions are somewhat lower.

Incumbents do not get blamed for droughts. Most voters willingly accept the sacrifices they must make when there are water shortages. Farmers who suffer crop damages can expect eventual insurance compensation from the US Department of Agriculture. Voters generally know and care more about what elected officials do during drought emergencies than what happens before and after. The public's interest drops as memories of a drought fade, making it hard to ask taxpayers to pay now to protect against future water scarcity.

This diminishing interest is widely acknowledged inside government circles. When Congress passed the National Drought Policy Act in 1998, it created a committee to explore ways address the imbalance between public underinvestment in planning, preparedness, and mitigation and the rising costs of emergency relief.[23] Accordingly, Congress put forward three core goals:

1. Favor preparedness over insurance, insurance over relief, and incentives over regulation.
2. Set research priorities based on the potential of research to reduce drought impacts.
3. Coordinate the delivery of federal services through collaboration and cooperation with nonfederal entities.

The committee's report noted that disaster relief had increased from $3.3 billion in the 1953–56 drought to $6 billion in 1988–89. Drought expenses across the United States now cost about $9 billion per year. The report notes that investing in drought preparedness is more economically efficient in the long run than paying for drought relief year after year, but the myopic practices continue. The report was replete with good suggestions about ways to prepare for drought but tactfully avoided criticizing government officials for repeatedly underinvesting in preparation.

One explanation for this derives from our earlier discussion about the federal government's general shift over time away from financing and building water infrastructure and toward environmental enforcement and the management of public lands. Instead of leading drought-resilience efforts, the federal government has defined its role as coordinating, collaborating, and supporting state and local governments. In the lingo of public management, the federal government would rather steer than direct water infrastructure development. The initial tasks of populating, defending, and developing western lands have been accomplished. Large water infrastructure projects such as dams and canals are expensive and highly controversial and therefore are more politically risky than rewarding. Deferring to the states and local communities on drought preparation also complies with US Supreme Court rulings on the principle of state sovereignty over inland waters.

Electoral incentives at all levels favor relief over preparation. Helping people in need is politically popular, whereas asking them to pay for protection against future drought risk is often a thankless chore. This applies across the board to many types of natural disaster relief, not just drought preparation. In an article appropriately titled "Myopic Voters and Natural Disaster Policy," Andrew Healy and Neil Malhotra try to account for why natural disaster relief spending in the United States increased thirteenfold since 1988, while money for preparation and mitigation has remained flat. They found that a doubling of relief payments before an election increased an incumbent president's vote share by a half a percentage point, whereas spending on disaster preparation had no electoral effects.[24] In other words, voters reward relief more than preparation. Healy and Malhotra point out that their finding fits into an extensive literature about voter myopia, which in turn shapes the incentives of elected officials. Voters prefer concrete benefits over collective goods. They also discount future benefits improperly and pay attention to problems that are primed toward media coverage. Politicians understand that giving voters what they want as opposed to what experts tell them they need is usually the better reelection strategy.

In short, Congress had constitutional, fiscal, and political reasons to cede drought preparation responsibility to state and local officials and limit its role to the more popular task of providing disaster relief. State and local government officials understand this dynamic as well. At a convening to share information and ideas for dealing with the prevailing drought conditions in 2015, western governors openly acknowledged that they could not count on the federal government "to develop water infrastructure," aside from some limited assistance programs such as offered by the Bureau of Reclamation's Water Smart program and the Environmental Protection Agency's revolving funds for infrastructure repair and modernization. Because "federal and state investment capacity is limited," local governments would have to "shoulder more of the infrastructure costs than they have in the past" through public-private partnerships and bond measures.[25] The job of building western water infrastructure has returned to where it began in the nineteenth century: the local communities.

Decentralized policies have predictable effects. On the plus side, they allow for more innovation and flexibility geared to the specific water resources and climate of a given area. But due to climate heterogeneity in the American West, some areas have water abundance while others are water poor. The combination of unequal water resources, legacy water rights, and a highly fractured water governance system inhibits an efficient and equitable sharing of water across local communities in the West. Even if that became the goal, it would be more difficult to build new large water infrastructure than it was in the past. Large-scale infrastructure projects in the past such as the California State Water Project and the All-American Canal transported water from plentiful to scarcer water areas. Attempts in the contemporary political era to mimic that strategy (e.g., the Peripheral Canal) have encountered severe opposition by environmental groups and residents in the water-abundant areas.

As a consequence, contemporary urban and suburban water strategies focus on smaller incremental approaches such as new local wastewater recycling, water efficiency measures, and matching grant incentives for local communities to engage in better regional planning and cooperation (e.g., California's Integrated Regional Water Management program). What these policies have in common is a lower price tag, less political resistance than building new dams or other large infrastructure, or both. This enables marginal progress on water supply but does not address deeper questions about water demands due to land-use policy (e.g., whether areas with depleted groundwater or that can expect cuts in surface water supply in the future should be allowing new agricultural or residential development).

Paying for these projects by bond measures rather than financing with general funds or new fees and taxes also makes water projects more acceptable to voters. Along the spectrum of western states, California has the reputation of being one of the bluest. One might assume that Californians are happy to tax themselves to provide for their water. In fact, Californians since the Proposition 13 tax revolt in 1978 have enacted a number of measures that make it harder for local governments to impose taxes without explicit voter approval and in some cases by supermajority thresholds. Bonds are less objectionable to voters. It is the equivalent of putting infrastructure expenses on a credit card as opposed to paying as you go.

Squeezed between the need for infrastructure and political obstacles to revenue enhancement, California relies heavily on bond measures for water infrastructure. Between 1970 and 2010, California voters passed twenty-one separate water bond measures worth a total of $23 billion. They covered a wide range of water needs including flooding, water quality, ecosystem restoration, recreation, and supply. Most are general obligation bonds that pay off the principal and interest payments over time. In effect, they share the fiscal pain with future taxpayers who will also benefit from water projects. Water bond measures almost always pass, but approval rates dropped from an average of 65% approval in the 1970s to 59% in the first decade of the new millennium.[26]

Spreading the benefits across diverse constituencies also increases the odds of passage. Usually, water bonds passed by the legislature with a two-thirds vote do not attract much controversy. Bonds put on the ballot by citizens are sometimes rejected when they seem to be skewed toward specific interests. An $8.9 billion water bond, for instance, failed in 2018 because it was perceived as favoring Central Valley agriculture excessively. But in general, bonds will likely continue to be the financial workaround of choice for states that cannot roll over debt in their budgets.

Another example of how water policy flows down the path of least resistance is the popularity of water conservation proposals. From 1980 to 2010, the California legislature passed eleven significant pieces of water legislation, five of which explicitly attempted to lower per capita consumption and encourage more efficient water use. Moreover, water conservation measures have been largely successful at achieving their goals. Per capita water use has dropped in California since 2005.[27]

While a sense of civic duty may play a role especially during a drought, water conservation policies have the political virtue of saving consumers' money. Such policies will please voters even though it hurts the profits of the utility companies. Rebates and reward programs are also popular. When the San Diego County Water Authority asked its customers what kind of water efficiency programs they were looking for, the top three responses were "rebates for water saving fixtures/appliances" (40%), "discounts for water efficient plants" (32%), and "rebates for water saving irrigation/landscaping" (30%).[28] People prefer to be materially rewarded for doing their civic duty.

The political reaction to building new water infrastructure for recycling or desalinating water is more mixed and not just because it costs taxpayers more money. Surveys indicate that desalination in particular is popular with California voters. However, environmental groups have concerns about the destructive impact of sucking in large quantities of saltwater and discharging high concentrations of brine back into the ocean. There are also public constraints on water recycling. A majority of Californians like the idea of using recycled water for outdoor watering but reject the idea of drinking or bathing with it.[29] This is despite the fact that Orange County has been putting recycled water back into the water supply for decades. Still, severe water shortages can be enough to overcome the public's aversions to recycled water. Los Angeles managed to pass a 2018 parcel tax with 69.5% of the vote that dedicated $300 million a year for treating stormwater and adding it back into the water supply. San Diego has also invested in both water recycling and desalination.

Underlying all water infrastructure choices are considerations of party and ideology. Even when water bonds are placed on the ballot by a bipartisan legislative vote, Republican voters are usually less enthusiastic about water infrastructure spending. All but one of California's most Democratic-leaning counties voted for Prop 1 (a $7.1 billion water bond), whereas half of the twenty most Republican counties voted against it. There was a similar partisan split in 2018 for Proposition 68 (a $4 billion bond for parks, the environment, and water).[30] Disagreements about the respective roles of the government versus the private sector can spill over into water policy.

CRISIS MANAGEMENT AND OPPORTUNITY

The notion that crisis creates opportunities for reform is a well-known political rule of thumb. In reality, whether this is true depends on the nature of the crisis conditions and the ability to move quickly when a window of policy opportunity opens. As discussed earlier, droughts vary in both length and intensity. If droughts are too short or not very intense, there will be little or no impetus for major investment in preparation, whereas severe drought emergencies lasting for many months soften the resistance to serious policy actions for a short period of time at least.

One popular version of this is the so-called windows paradigm.[31] Its basic premise is that major policy shifts require a convergence of three conditions: problem definition, policy development, and the political prospects for passage. The first condition requires that voters and public officials have to recognize that a policy problem needs a solution. It is very difficult to allocate scarce resources for a water project unless people view it as a pressing need. This tends to wax and wane with the alternation of the drought cycle. The second condition is the development of possible solutions. In the case of drought management, it could mean adopting technical innovations (e.g., more water-efficient appliances, better or irrigation systems), shifting water infrastructure strategies to avoid strong environmental objections (e.g., from dams and reservoirs to aquifer replenishment), or altering water allocations between farms, fish, and people. Finally, the political conditions have to be right to put together a coalition that is strong and durable enough to break through the systemic inertia created by water rights, governance fracture, interest group politics, partisanship, and voter hesitancy to pay more taxes. The policy windows theory suggests that emergency situations such as serious water shortages can make possible what is normally politically blocked.

There is some evidence that severe drought conditions do indeed open up a brief window for water policy change. California's most recent severe drought began in 2012–13. By 2014, Governor Brown issued his first drought emergency proclamation. Citing the previous two years of low precipitation, a forecast of an even drier year in 2014, and the fact that the snowpack was 20% below normal, the governor called upon the state to undertake a 20% reduction in water consumption.

The governor's initial foray into handling the drought was respectful of California's strong traditions of home rule and local government's resistance to heavy-handed state intervention into water management. The governor called on local municipalities and water agencies to invoke their drought management plans but did not institute mandatory cuts. While this might on the face of it seem like a rather weak intervention, research has shown that it actually resulted in a relatively high level of voluntary compliance due to extensive media coverage in the preceding months. Matching water data trends with the intensity of media coverage, Quesnel and Ajami show that an increase of one hundred drought-related articles in a bimonthly period resulted in an 11–18% reduction in water use.[32] Various incentive programs and differential pricing also played an important role, but to a considerable degree much of the water savings was due to voluntary compliance of people wanting to do their part in an emergency or at least paying heed to social pressure from the neighbors. However, voluntary actions are often not enough, and as conditions continued to deteriorate, the governor issued an executive action in 2015 that upped the goal of water reductions to 25% and directed the State Water Resources Control board to impose restrictions that would meet that goal.

The main goal during a drought crisis is to achieve higher levels of water conservation. Setting a statewide goal can be politically complex in a state that has many different climate conditions and where past conservation efforts have varied so much. Water use along the milder and cooler coastline is substantially lower than in the hotter, drier inland areas of California. Some areas of the state such as Los Angeles had experienced more droughts than in the northern parts, and had already taken more serious steps to reduce water consumption. While the governor's formula did not give the diligent communities enough recognition for the water savings they had already achieved prior to the emergency proclamation, the communities in southern California did not challenge the guidelines in any serious way.

Utilizing the framework of information, social conformity, and carrots and sticks as possible management tools, we can see the following. Information about water scarcity was widely disseminated by the press and made people aware of the drought and the need to conserve water. Policing drought policies would require far more manpower than most localities had, but social pressures disposed people to comply. Some referred to it as drought snitching. Neighbors reported water wasters to Los Angeles County's Water Conservation Response Unit, which then triggered a warning letter about excessive sprinkler use and hosing down a sidewalk.[33] Violations can lead to "sticks" in the form of fines, which are generally not popular with voters. When people do not like policies that cost them more money, they sue. Lawsuits introduce judicial uncertainty into the policy calculation. Such was the case with tiered pricing schemes that some water utilities resorted to that charged higher prices for heavy water users. But in the end, people responded well to the water crisis. Californians actually cut back more than the 25% mandatory level.

Unfortunately, what is done during an emergency period does not necessarily carry over to recovery and preparation. When the drought proclamation ended, the mandatory cuts were abolished. The governor had used his emergency powers to temporarily suspend the state's environmental review law, the California Environmental Quality Act, in order to facilitate projects that could enhance the water supply, but that too ended in 2016. Moreover, while some incremental water efficiencies persisted, studies using smart water meters reveal that many consumers slid back to higher water use after the social pressures abated and the restrictions lifted.[34] Net of the reversions, there was still incremental progress but not the steady progress or significant reform that would be the best form of climate adaptation.

———

Returning to the question of whether a drought crisis can lead to serious structural policy change, we can note the following. First and most importantly, the drought spurred the state to finally regulate ground-water use. Unlike many other western states, California had for many decades regulated surface water but not groundwater.[35] Farmers and others who were fortunate to own land above aquifers were able to draw groundwater to compensate for when they could not get their full quota of surface water. Over time, aquifers throughout the state but especially in the Central Valley have been depleted, and land has subsided as the water has been withdrawn, causing infrastructure damage. The new groundwater law designated areas where the groundwater depletion was most severe and demanded that new agencies form and create groundwater sustainability plans.

Second, although many consumers did slide back to higher levels of water use after the emergency quotas were lifted, there is evidence that some of the water measures, such as low-flow water appliances and more drought-resistant plants, were retained. This has added to the achievement of ever-lower per capita water use in California.

That said, there is still the question of how the state will handle future, more intense, droughts without addressing underlying structural prob-lems, particularly if water shortages occur more frequently or are more intense. At some point, the trade-off between environmental flows, agricultural use, and urban-suburban residential and commercial use could become unmanageable, causing the entire system with its inherited legacies to fail.

CHAPTER 3

WILDFIRE TERROR AND POLICY SPARKS

Wildfires are a vivid reminder that the American West is still a risky place to inhabit, particularly if you are located in or near wildland areas. Due to the seasonality of precipitation in many areas of the West, vegetation flourishes in the winter and dries out in the summer, creating the fuel for fires in the fall. Fanned by strong winds blowing from warmer inland areas, ignitions can grow rapidly into major firestorms, causing deaths and extensive destruction. Due to global warming and more severe droughts, vegetation is drying more extensively, increasing the potential fuel load for future wildfires.

However, wildfires are not purely natural occurrences. They are usually ignited by human activity, which means that to a greater degree than with most natural disasters, wildfires are both a people- and a nature-based problem. As populations expand into wildland regions in search of either cheaper housing or beautiful natural surroundings, the ignition risk goes up. Keeping populations out of these areas would lower the odds of starting fires and mitigate the property damage from the flames. However, the quest for new revenues from property development and the availability of wildfire insurance incentivizes local communities to build—and rebuild if necessary—in wildfire-prone areas. This, in turn, increases the odds of future wildfire ignitions. In this way, the cycle of wildfire vulnerability builds on itself.

Unlike the relatively long ramp-up to a drought crisis, wildfires arise suddenly. Fleeing a rapidly approaching wildfire is very traumatic. After

experiencing a 2020 dry lightning storm that ignited 650 separate wildfires in northern California, one evacuee observed wryly that she was
thinking of moving to Florida because at least there "you have several
days to plan for a hurricane," whereas "for fires, you're woken up in the
middle of the night."[1] While most wildfire incidents are put out rapidly,
it can take weeks or months to completely extinguish the largest ones
in forested areas. The manpower and costs needed to deal with large
wildfires are massive and likely to escalate in the future.

Wildfires also generate giant smoke plumes that can travel long distances in the wind, fouling regional air and affecting millions in very
unhealthy ways. The more extensive reach of wildfire smoke as compared
to the danger from the flames per se is clearly evident in figures 3.1a–b.
When asked in the Bill Lane Center for the American West's 2019 western
regional survey whether they knew if their state had experienced any
wildfire event in the previous twelve months, more than half of California
and a third of all western residents said that they knew it had. But only
13% of Californians and 8% of all western residents claimed that they had
personally experienced a wildfire during the previous twelve months,
and more than half (52%) said that they had been subjected to wildfire
smoke during that time. The exposure to both flames and smoke was
higher in 2020, but the same pattern of higher exposure to smoke than

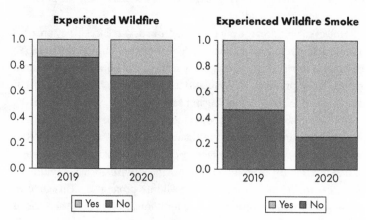

Figure 3.1. Proportion of Californians in the Bill Lane Center's 2019 western
regional survey and 2020 California wildfire survey who had and had not
personally experienced a wildfire (*left*) or wildfire smoke (*right*).

to flames was also evident. After many weeks of being enveloped in smoke, one weary Bay Area resident observed that she and others "are tired of smoke. We've had smoke overlaying us for the last three weeks, and the skies are just beginning to clear."[2]

Wildfires illustrate some important aspects of how citizens view risk and responsibility in choosing where to live. Two key elements of extreme weather risk are the lethality and extent of exposure for people, infrastructure, and property. Wildfire flames are highly hazardous, particularly when winds rapidly spread sparks and flames across a wide area. Given the high degree of urban density in the West, most people do not live in or near the high-hazard wildfire areas. Wildfire flames are obviously highly lethal, but the exposure to them is mostly limited to rural populations and the edges of exurbs. Wildfire smoke, on the other hand, can expose large numbers of people to health dangers. The political impact of wildfire smoke could be enormous, but that is not the case presently because most people are unaware of the health risks associated with prolonged wildfire smoke exposure. In many ways, it is a situation analogous to the public's lack of awareness about secondhand tobacco smoke until the 1980s.

Despite the risks, people continue to move into dangerous wildland areas, which increases the odds of fire ignitions and adds toxic chemicals to the smoke plume when structures burn. Underneath the inability to control residential and commercial growth in wildland areas is a deeply embedded western conviction that people have an individual right to live in risky settings even as the social costs of fighting and recovering from wildfires continue to soar.

WILDFIRE AS BOTH A NATURE
AND A PEOPLE PROBLEM

Wildfires are a nature problem in the sense that they are a product of the more extensive droughts and distinctive weather seasonality in many parts of the American West. Seasonality, as discussed earlier, refers to the annual cycle of rainy winters that yield new vegetation that dries up during the summer months and becomes wildfire fuel in the fall. The West's distinctive topography also contributes to its wildfire problem. Winds from the inland areas pick up speed and grow warmer as they move up and down the slopes and through the gaps in the mountain ranges. The so-called Diablo and Santa Ana winds in the far western states can travel at speeds of over forty miles per hour, toppling power lines and spreading fire sparks at high velocity.

The most destructive wildfires are crown fires that engulf entire trees and spread rapidly with the strong winds. Large trees can burn for many hours and produce a great deal of heat. Rangeland brush fires, by comparison, burn more quickly and produce less heat but can be very destructive as well. The aftermath of major wildfire events results in burned landscapes that are conducive to flash floods and debris flows when the winter rains return. And to add more misery to the equation, megawildfires release large amounts of carbon dioxide into the atmosphere, partially wiping out efforts to reduce emissions and contributing to further global warming.[3]

The core management dilemma is that while wildfires have all of these adverse impacts, they also promote healthier forests. Turning dead trees and plant matter into ash returns nutrients to the soil. Fires also clear a path for sunlight to the forest floor and destroy invasive weeds and insects, fostering the growth of native species and preventing a larger fuel buildup. This lowers the intensity of subsequent conflagrations.[4] It is quite likely that the US policy of aggressive fire suppression throughout much of the twentieth century worsened the current situation by allowing the buildup of potential fuels in forested areas. Modern forestry practices of prescriptive burns and mechanical thinning could alleviate this accumulated problem but run into substantial timing, expense, and permitting issues.

There are inherent trade-offs in fire policy decisions. Suppress wildfires too much and you lose their natural ecological benefits and perversely provide more fuel for future fires. Suppress too little and the fires can cause extensive property damage and endanger lives. Forest fires in very remote places threaten fewer lives and property but nonetheless discharge large plumes of toxic smoke over wide areas of the American West. Letting fires in remote locations burn themselves out cleanses the forests of underbrush in a natural way but at the cost of emitting smoke and harmful particulate matter into the air.

One commonly used measurement of the amount of particulate matter in wildfire smoke is an air quality index that ranges from 0 to

500. Wildfire smoke can contain many harmful substances, not just particulate matter from wood and vegetation but also toxic materials from burning structures and vehicles. Inhaling unhealthy air (i.e., an air quality index value above 200) for prolonged periods of time or while exercising vigorously can be the equivalent of smoking multiple packs of cigarettes. Those who live closer to the fires or those fighting them can be exposed to hazardous air quality index values in the 400–500 range. Serious wildfire smoke events lead to higher rates of hospitalizations for asthma, bronchitis, and heart problems and ultimately higher rates of mortality and a weakening of the body's immune system.[5]

Even small-scale prescriptive burns that are meant to limit or prevent larger uncontrollable wildfires emit potentially harmful PM2.5, fine inhalable particles that are generally 2.5 micrometers and smaller in diameter, albeit at lower levels. As those who apply for permission to do prescriptive burns often discover, neighbors in many instances oppose planned burns that are intended for their own protection due to their concerns about smoke or the remote chance that a controlled burn might become uncontrollable.

Adding to the nature challenge is the likelihood that global warming will increase the frequency and intensity of western wildfires in the future. If temperatures continue to rise, the number of consecutive dry days will increase, and the soil moisture content will decrease. The dried vegetation will become more susceptible to combustion, leading to longer and more intense wildfire events.

Efforts at climate change mitigation and wildfire resilience are inextricably linked: it is not a choice between one or the other. The costs associated with preventing, fighting, and recovering from wildfires will necessarily compete with the resources needed to decarbonize the grid and the transportation sectors, but without decarbonization, future fires will be both harder to fight effectively and more expensive.

People exacerbate the future wildfire risk as more of them move into or near wildfire-prone areas. In some cases such as Paradise, California, remote rural areas attract retirees and others who cannot afford the rents and mortgages of urban and suburban homes. But the wilderness appeal also draws people of means who aspire to inhabit a ranchette in an "amenity West" setting. In other words, people along the entire socioeconomic spectrum contribute to population growth in wildland-urban interface (WUI) areas.

WUI areas are defined as those where human structures and infrastructures exist within or adjacent to wildfire-prone areas. They can be found throughout the country and expanded from 7.2% to 9.5% of total US landmass between 1990 and 2010.[6] Westerners moving into these places are usually aware of the risks associated with locating in WUI lands because in many instances there have previously been fires in these places. Sixty-one percent of new WUI homes in the last two decades were built in areas that had already been classified in 1990 as being prone to wildfires. People even rebuild in areas that had burned in recent decades. Housing units built in previous fire areas increased from 177,000 to 286,000 units between 1990 and 2010.[7] As people and infrastructure intrude into these spaces, the odds of wildfire ignition as the result of human actions also go up. The rain may not follow the plow, as early settlers apparently thought, but wildfires follow the people. Over 80% of wildfires are caused by humans and range from sparks emanating from lawnmowers and downed electricity transmission lines to arson.

LEGACIES OF THE PAST POLICIES AND INSTITUTIONS

As noted several times, past policies have lingering effects. In the case of wildfires, earlier decisions to aggressively suppress fires now contribute to the severity of the contemporary forest management problem. For much of the twentieth century, US policy was to put wildfires out as quickly and assiduously as possible. The extensive destruction of a 1910 wildfire that burned three million acres in Idaho and Montana convinced the US Forest Service that fire suppression had to be a top priority. Under the leadership of Ferdinand A. Silcox, the Forest Service aimed at putting out fires that started on one day by 10 a.m. the next day. The federal government's suppression strategy was later extended to state and private lands through the Clarke-McNary Act of 1924.[8]

Part of the reason that aggressive fire prevention was so widely and readily accepted is that it was very much consistent with the federal government's general mission in the nineteenth century through the mid-twentieth century to develop and utilize western natural resources for commercial purposes to the fullest extent. Fire suppression not only saved lives and property but also protected the lumber that could be harvested and sold by the timber industry.

Little or no thought at the time was given to ecological concerns about the role that fires played in maintaining healthy forests, even in academic circles. Early scholarly advocates for preventive prescriptive burning such as Harold Biswell were not initially taken seriously by other scholars and government officials, in part because the orientation of forestry and natural resource departments in many state universities at the time leaned also toward the commercial utilization of forests, not their ecological benefits.[9]

This lack of ecological concern also reflected the influence of timber and agriculture interests on malapportioned state legislatures prior to the US Supreme Court's decision in *Baker v. Carr* (1962). In addition, lobbying and campaign finance laws were much weaker in this period. This changed after the post-1970 political reforms. Firefighting doctrine changed in forestry schools and departments. Gradually, the view that fire suppression interfered with the natural processes for removing excessive fuel buildup and made wildfires ever more dangerous over time took hold. But it is a case of too little too late for many forested areas in the West, given the high fuel loads that have accumulated over many years of aggressive fire suppression.

The fact that the federal government swung from assiduous fire suppression to the modern "let it burn" strategy in a relatively short period of time may have weakened to some degree the public's confidence in government fire policy, especially for those who grew up in the era of Smokey the Bear commercials and for longtime residents in the WUI who had to adjust to the new policy. It is one thing politically to build upon and refine previous scientific judgment; incremental change is consistent with the public's general conception of progress. But, it is quite another matter to reverse course entirely, especially when the new approach potentially threatens the health and livelihoods of those who work, live, and recreate in the forests and rangeland areas. While contemporary forestry science almost unanimously supports the practice of prescriptive burns and letting wildfires in remote places run their course, it is not surprising that these policies remain controversial for many WUI residents.

Another example of the American West's legacy shaping current wildfire policy is the area's highly fractured governance space. To be sure, fracture is an omnipresent feature of American federalism and, as discussed earlier, complicates water management and drought resilience as well. An important difference, however, is that western water policy is largely under state and local jurisdiction, whereas forest and rangeland management is divided between many state and federal agencies. The federal government owns and manages a great deal of western land. In California, for instance, it is responsible for 46% of the state's total area versus 31% for the state and 23% for local governments. Coordinating with the federal government on wildfire policy is therefore essential.

But within the federal government, six different agencies (the US Fish and Wildlife Service, the Bureau of Land Management, the National Park Service, the Bureau of Indian Affairs, the Forest Service, and the Department of Defense) also share responsibilities for the various types of federal lands. This creates both vertical and horizontal coordination problems, vertical in the sense that local and state governments and the federal government must coordinate with one another and horizontal in the sense that various agencies at each level are involved. Federal agencies are well aware of the problem and have attempted to solve it by creating a common framework under the Federal Wildland Fire Management Policy, agreeing to coordinate with each other regardless of which US agency owns the land a wildfire starts on. Other examples of federal firefighting coordination efforts include the National Interagency Fire System and the National Wildfire Coordinating Group.

Governmental fracture is also a problem below the federal level. State governments have to deal with many different types of local entities within their boundaries, including cities, towns, and various kinds of special districts. Many of these local entities also have to collaborate with one another to achieve any kind of coordinated wildfire policy at subregional and regional levels. But localities in general jealously guard their control over residential and commercial zoning and development. The economic incentive to generate more revenue to pay for municipal services drives communities to build new shopping centers or residential properties near or in WUI areas even though such locales entail more wildfire risk.

State and federal programs can also try to encourage and materially incentivize local vegetation management and home-hardening programs, but without serious local engagement, they will not succeed. Localism—the preference for maintaining local control over resources and decisions—frequently undermines efforts at regional and subregional fire management collaboration even during emergencies. For instance, although California cities had signed mutual aid agreements to lend each other equipment and firefighters in times of emergency, many of these requests for local aid during 2018 wildfires were unmet because towns had to weigh whether to lend their resources or keep them in the community in case they were needed to fight fire outbreaks in their local area.[10]

Governmental fracture has advantages and disadvantages. For public services that can be provided efficiently at the local level with little or no spillover effects, fracture can theoretically provide consumer choice, allowing people to vote with their feet for the particular bundle of public goods that they want and are willing to pay for. But when spillover effects and collective problems require collaboration, localism leads to ineffectiveness and inefficiency.[11]

Wildfires have considerable spillover effects. If a jurisdiction in or adjacent to a WUI area does not institute proper vegetation management or if private land owners do not undertake prescriptive burns or mechanical thinning of their forest areas, fires that ignite on the unmaintained land can spill over into urban and suburban neighborhoods many miles away. And as noted before, harmful smoke can migrate even farther, causing both acute and chronic health problems. Shared wildfire policies require novel jurisdictional collaborations as well.

WILDFIRE POLICIES

What, then, are the policy options and legacy political constraints that western communities face today when managing and dealing with wildfires? As noted earlier, wildfires constitute one of the gravest threats to life and property across the spectrum of extreme weather events, capable of igniting suddenly and spreading extensive amounts of smoke that cause serious short- and long-term health consequences. To put it into the language of risk assessment, the hazard is high, and the exposure of people and property can be extensive. And the vulnerability of western communities to large destructive wildfires is likely to get worse under conditions of future global warming.

As with extreme weather policy generally, the elements of wildfire management are prevention, emergency operations, and recovery. Prevention, as noted before, is the most efficient way to handle extreme weather risk because the savings from preventing ignition and mitigating damage can offset the escalating costs of firefighting and recovery efforts, but political incentives tend to favor post disaster relief over taking protective steps. It is easier for a politician to vote for emergency funds in the immediate wake of a natural weather disaster when sympathy for the victims is at its peak than to ask voters before a disaster or many years after (when public attention and memory have decayed) to materially support measures to protect people from something that might hypothetically occur in the future.

According to policy windows theory, the direness of future wildfire scenarios as compared to other kinds of weather-related events should weaken the various inertial political, psychological, and social forces that typically undermine support for adopting protective measures. Problem salience—that is, people recognizing that there is a serious problem to be solved—is a necessary but not sufficient condition for collective action. Voters and taxpayers also need to believe that a problem has widespread social consequences for them and others that requires collective solutions. Otherwise, they will regard a risky choice as an individual problem without serious social consequences.

Wildfires have in recent years certainly achieved sufficient public salience throughout the American West and in the national media to create many windows of opportunity for significant policy changes. People living in the region have been inundated with horrific images and extensive media reports about wildfire events. Residents are generally aware that wildfires can be extremely dangerous. They also tend to have great sympathy for wildfire victims as well as strong feelings of concern for themselves. When the Bill Lane Center's western regional survey respondents were shown pictures of recent wildfire damage, the top reactions were sadness (57%), worry (55%), anxiety (34%), and fear (32%). Even so, a majority of them still framed the problem as one of individual choice and responsibility for the risks people assume when they decide to live in wildfire-prone areas, ignoring the social costs associated with fighting fires and helping communities recover and rebuild. This puts severe constraints on the steps that communities are willing to take to protect themselves from future wildfires.

PREVENTIVE MEASURES

The central resilience choice is between making people and property safer in risky places versus making them safer by keeping them out of risky places. The prevailing public opinion in the American West favors keeping people in place and encouraging but not requiring them to take steps to make their situations marginally less dangerous. But here's the rub: the public also generally supports large relief packages for communities that experience damaging wildfires. In other words, westerners believe that individuals should be free to live in dangerous areas if they so choose but also believe that the government should bail them out with emergency support and relief services when their bets with nature go awry. In so doing, insurance creates an unintended moral hazard problem: providing insurance against risk encourages more risky behavior.

This confused and shortsighted thinking is not restricted to the American West by any means. The same can be said of midwesterners who live in river flood plains and Miami residents who live in coastal properties prone to both hurricanes and flooding due to sea level rise. Giving relief to people who make risky choices encourages them to do more of what we do not want them to do: live in dangerous areas. Government-subsidized insurance programs and publicly funded wildfire fighting services in effect encourage riskier locational choices by enabling people (or at least seeming to do so since homeowners are often underinsured) to rebuild if their home or commercial property is destroyed by natural disasters. While such assurances are compassionate and popular, publicly funded safety nets induce more people to move into or continue to live in dangerous wildfire areas even as the costs of putting out, recovering from, and paying for the liability for fires escalate. Just as the US government encouraged people to move into the often dangerous western frontier by offering them land deals and building water infrastructure, modern policy encourages people to live in high-hazard wildfire areas by providing firefighting services, requiring connection to the electricity grid (even though downed lines in wooded areas cause many wildfires), and providing insurance of last resort and guidelines for creating defensible spaces (which actually are at best partially protective).

Private insurance at least has the offsetting feature that companies may eventually raise their rates so high to compensate for huge losses that it will send market signals that living in wildfire-prone areas should be discouraged. A private insurance market can only flourish on its own if the risk of some extreme weather event is low enough such that a company can make a profit from its premiums even as it compensates homeowners who sustain a loss.

As megawildfires have become more frequent and dangerous, payouts have escalated and premiums have risen. Some insurers have become reluctant to insure homes in fire-prone areas. After two bad wildfire seasons in 2017 and 2018, the insurance companies doing business in California had $24 billion in losses. The Camp Fire in Paradise alone created so much damage that it caused the Merced Property and Casualty Company to fail, prompting the state's insurance commissioner to fret that California was "slowly marching toward a world that is uninsurable."[12] When homeowners had trouble getting their insurance companies to renew their policies, the state legislature stepped in and banned them from refusing to renew fire policies for one year after a wildfire disaster.

Moreover, when Californians cannot procure standard property insurance due to high wildfire risk, they can turn to the insurer of last resort, the California FAIR plan. Although that plan is capped in payout, is more expensive to purchase, and does not include standard homeowner liability or theft coverage, it nonetheless helps people rebuild when they are burned out.[13]

When westerners were asked whether residential property owners in wildfire-prone areas should be required or merely allowed to buy wildfire insurance, a majority preferred that homeowners should have the option but not the requirement to buy it. This was a little less true of Democrats (45% optional to 41% required) than Republicans (59% to 30%), and Independents (54% to 34%), reflecting to a modest degree the ideological divide over climate change associated with partisanship. Although the private wildfire insurance option is widely supported, there was no majority support among the survey respondents for helping residents with public subsidies in wildfire-prone areas except in the case of the truly disadvantaged. Democrats were a bit more inclined than Republicans and Independents to support the idea, but still not a majority of them thought so.

Pretty much the same individual responsibility logic applies to home hardening, vegetation management, and other steps to upgrade a residence in a wildfire area in order to make it more resilient. While it is nearly impossible to prevent a home from being destroyed by the most intense wildfires, it is feasible to make homes more resistant to them by taking such steps as replacing open eaves with closed ones, substituting wood shingle siding with fiber cement board, and using rocks as mulch rather than wood chips or bark. But all of these retrofits can be very expensive, particularly if rooftops are involved. And in the end, the heat and torrent of sparks will prevail if they are intense enough or if people are forced out of their homes so quickly that they do not have time to close off the vents or close the garage during their scramble to safety.

Much the same could be said of efforts to create defensible space around a home through vegetation management. This involves creating perimeters around the house and other nearby structures that are free of flammable vegetation and having only very limited vegetation within one hundred feet of the structures. Unfortunately, removing lush vegetation often diminishes the aesthetic appeal of a home in nature, which discourages homeowners from taking steps that would protect them. While westerners in our study were a little more disposed to using public funds to encourage homeowners to make upgrades that would better protect against wildfire damage (41%, including 52% of Democrats), many were still either opposed to the idea or did not have a view. And in the end, these steps only lower the odds of damage. There is no iron-clad guarantee of protection from destruction.

Other preventive actions offer more protection to the community as whole, such as prescriptive burns in the surrounding WUI areas. These are controlled burns before the fire season starts that eliminate the ladder fuels on the ground that enable flames to start their climb to the treetops, leading to hotter and more dangerous fires. Westerners favor requiring private owners and communities to do prescriptive burns (49% in favor, 33%, opposed), including pluralities of Democratics, Republicans, and Independents. But there is no strong majority. This matters politically because without a requirement to do prescriptive burns, private owners are understandably reluctant to deal with the burden and costs and the regulatory steps that are involved in getting permission. There is also the small but highly problematic possibility that controlled burns can occasionally become uncontrolled burns, unintentionally inflicting the very damage that the property owner was trying to avoid.

Many experts believe that given the projections of global warming, the most efficient policy solution for minimizing wildfire risk would be to use zoning laws to discourage or prohibit individuals and commercial establishments from living in high-hazard wildfire areas. Insurance, as previously discussed, is a soft form of this tactic. Residents and business owners in an area that is zoned as a high wildfire risk will likely face higher fire insurance rates. This has the desired effect of discouraging some would-be residents from living in a wildfire-prone area. On the other hand, linking higher insurance rates and more stringent regulatory scrutiny to particular zones can be politically controversial. Existing residents and business owners tend to resist paying for more expensive insurance costs, and developers will complain that it discourages future construction. This tension can turn zoning meetings into heated battle grounds.

California established its Fire Hazard Severity Zone program in the wake of a fire in Panorama in 1980. This designation was based on such factors as the fuel characteristics, land slope, and local weather. Still, it is almost impossible to predict whether a fire will happen in a given place at a given point in time. The best estimates express the likelihood of a fire occurring in an area within a range of time such as thirty to fifty years. But the underlying uncertainty at the smallest scale about where and when a wildfire might occur fuels the inclination to contest the insurance rate line-drawing process. Consumers understandably resist dramatic increases in insurance costs no matter the rationale. And given continued global warming, the dangerous zones will increase in size, which will mean more consumers facing higher insurance rates. Specific determinations of high-hazard wildfire zones often seem to neighbors as arbitrary, especially since major wildfires do not confine themselves to designated places. Homes that are adjacent to but not included in high-hazard wildfire zones can still be highly vulnerable to out-of-control wildfires even though they are not subjected to the same higher insurance or regulatory burdens.

The insurance problem, in short, is very difficult. On the one hand, allowing people to have wildfire insurance coverage encourages risky location decisions, but on the other hand, the absence of insurance complicates the problems of recovery. Higher wildfire insurance rates can also cause more residents to underinsure or forgo wildfire insurance altogether, which then leads to a higher burden of homelessness in burned-out areas.

The 2017–18 fires in northern California forced many people from their destroyed homes into camps and shelters in the areas around Paradise and Santa Rosa. Those who were underinsured or lacked insurance remained essentially homeless for extended periods of time. Many of them were retirees or disadvantaged families with limited resources and savings to fall back on. This in turn increased the social service burden on these impacted local communities at a time when they were experiencing revenue loss from diminished property and sales taxes due to fire damage. Sympathy for the victims and the fiscal urgency to restore lost revenues caused local officials to expedite the process of rebuilding, perpetuating the cycle of wildfire risk.

The toughest preventive step would be for local governments to prohibit new development in wildfire-prone areas, but this is difficult for communities to do for economic and political reasons. Local governments depend on commercial and residential growth to provide revenues to pay for the public services they provide and to support their active and retired personnel. New properties provide additional property tax revenue, and new businesses add sales tax revenue. If properties are destroyed by fire, the community loses tax revenues at the same time that it has to pay for emergency services and cleanup. Some of this can be offset for a short period of time by federal, state, and private emergency relief money, but unless properties are rebuilt, the local community faces the prospect of long-term revenue loss.

From an outsider's perspective, it may seem perplexing and foolish for local governments to allow citizens to rebuild in high-hazard fire zones, especially those that have recently experienced wildfires. But from the perspective of restoring lost revenue and responding to public sympathy for fire victims, it is not surprising that so many communities follow this rebuilding strategy.

Looking at the 2019 western regional opinion survey we conducted, we can see that public sentiment is mostly wary of strong zoning laws that would limit the ability of existing homeowners to rebuild in WUI areas. A majority of westerners in the regional survey signified their support for allowing property owners to rebuild in areas that were destroyed by wildfires as opposed to forbidding them to so. A variation on the not rebuilding theme is strategic retreat. As applied to wildfires, this would mean relocating individuals out of high fire-hazard severity zones into safer spaces. But westerners mostly opposed mandating a policy of relocation and favored letting people decide for themselves.

Potentially important exceptions to this laissez-faire strategy are requiring prescriptive burns and restricting new residential and commercial growth in dangerous wildfire areas. A majority were in favor of new growth restrictions as opposed to forcing retreat or preventing existing residents and businesses from rebuilding in wildfire-prone areas. There seems to be an important difference in the public's mind between taking actions before people and businesses have sunk costs into already-existing property versus letting people move into areas of high wildfire risk. This suggests that proactive steps before development goes forward in the WUI areas is more politically feasible than imposing costly actions on current residents. But a glimmer of public support does not guarantee smooth sailing by any means. Lobbying by local developers will be fierce, and local officials will be mindful of the local revenue imperative as they consider WUI growth.

Short of forcing people into strategic retreat, it is also possible to use public funds to incentivize them to do so. Here the picture of public support is murkier. The least popular options—opposed by the majority of voters—involve using public money to buy out residential and commercial property generally without taking consideration of family income. There was, however, plurality support for using public subsidies to encourage low-income families to relocate or to buy out their properties in order to get them out of harm's way. But as we have seen before, proposing the use public funds for these purposes tends to divide voters along partisan lines.

In sum, westerners favor allowing owners to stay in place and protect themselves with home upgrades and buying insurance. Westerners are open to the idea of not making a bad situation worse by limiting new growth in wildfire-prone areas but respect the right of people who already live in these areas to remain there. With the exception of low-income families, they believe that the risks and expenses of living in high-hazard areas should largely be borne by the private owners, not the public sector. It is unclear whether many understand the collective costs and health effects of wildfire smoke and fires that are associated with letting individual homeowners live in or next to high-hazard wildfire areas.

THE URGENCY AND DANGER OF WILDFIRE EMERGENCIES

Drought policies typically involve the curtailment of water usage until a low-precipitation period ends. While water restrictions can force people to take shorter showers or let their lawns turn brown, modern American droughts rarely result in major population displacements or water-deprivation deaths. There is not the same urgency to act quickly in a drought as there is in a major wildfire. Drought officials have days and weeks to deliberate over various options about water emergencies without any immediate concern for lives hanging in the balance if they fail to act rapidly. Wildfires, by comparison, are extremely dangerous and fast-moving events, often requiring critical life-and-death decisions in the moment by residents and firefighters to evade the unpredictable path of the flames. Ideology and party loyalty take a back seat to survival when intense fire, heat, and smoke are coming at you at high speed.

At the same time, there are implicit assumptions behind those momentary decisions about where to deploy fire crews or what to save from fire damage. Western forests benefit from periodic burns. Without the people and valuable structures in wildfire-prone areas, the best forest management policy might dictate to "let it burn" in order to eliminate ladder fuels expediently. But human settlement and economic development require protection, which means that the firefighters' top priorities are saving lives and protecting property.

As discussed earlier, the goal of suppressing fires as rapidly as possible has a long lineage. Only in recent decades has it yielded to a greater appreciation for the salutary benefits of controlled burns. Except in the most remote areas, uncontrolled burns are simply too dangerous to the people and property in wildland areas to be ignored. Given the projections of future population growth and economic development inside and adjacent to WUI lands, the demand to contain wildland fires will only grow.

Fighting wildfires is dangerous and expensive. While there have been and will continue to be improvements in the technology for detecting and combating flames, the unpredictable direction of wildfires and the intensity of their heat make wildland firefighting a hazardous business. As recently as 2013, 19 firefighters died fighting a fire in Yarnell, Arizona. Since 1990, 480 firefighters have died during wildland fire operations, including 170 between 2009 and 2016.[14] The economic costs of fighting fires are also high and rising for both the states and the federal government. California spends about $2.5 billion per year on firefighting. The federal government's national wildfire-suppression costs have risen steadily since 1985 from $200 million to $3 billion per year.[15]

Again, wildfires are as much a people as a nature problem, and both are getting worse. Global warming will lead to more consecutive dry days, decreasing soil moisture and producing more dried-out vegetation. This will create the conditions for longer fire seasons and more extensive and severe wildfire events. At the same time, the continuation of people moving into wildland areas will increase potential property loss and deaths and greatly magnify the firefighting burden.

These impacts are especially true for exurban and suburban sprawl as opposed to isolated cabins and ranches. It is easier to protect one structure in a sparsely populated area than a cluster of homes that are inside or adjacent to wildland forests and rangeland chaparral. A study by Headwaters Economics, a nonprofit research group focused on improving community development and land management decisions, concluded that firefighting costs in Montana were highly correlated with the number of homes located in proximity to a wildfire. The group estimates that for every 150 homes threatened by fire costs, the State of Montana would incur an additional $13 million in suppression costs, and if current development trends continued, wildfire costs could double between 2006 and 2025.[16]

When lives and properties are lost, there are often associated liability costs. Most fires are human-caused in various ways. One common source, for instance, is ignition by sparks from downed electricity transmission lines. Unless the electricity lines are buried underground, which is expensive and not feasible in certain types of terrain, strong winds can topple the poles or cause the lines to detach. Seven out of the top twenty most destructive wildfires in California's history were caused by downed power lines as compared to just one by lightning and two by arson.[17]

In the case of California's Camp Fire in 2018, PG&E's equipment was determined to be responsible for an ignition that resulted in eighty-six deaths and destroyed 13,600 homes. Under California's doctrine of inverse condemnation, the company was held strictly liable for the damages. After lengthy bankruptcy proceedings, PG&E agreed to pay $24.5 billion in damages. In addition, it had to come up with money to harden its grid and improve its vegetation management around the power lines. Utility companies pass off their liability expenses to consumers in the form of rate increases. Rate increases can make it harder to ask consumers to also pony up for other climate change policies such as building new energy storage capacity or more utility-scale energy sites. Wildfire prevention costs also divert funds from a state's decarbonization commitments. After the fires in 2019, California had to set aside $170 million from its cap-and-trade proceeds to pay for wildfire prevention measures. In a world of limited tax resources, resilience now competes with greenhouse gas mitigation projects for public funds.

Aside from the danger of destruction from the flames, wildfire emergencies also generate enormous amounts of toxic smoke across an extensive area. From a political point of view, the smoke from wildfires has a wider potential political impact than the spread of flames. The West is heavily urbanized, which means that many westerners live in densely settled urban and suburban clusters that are far removed from the flaming embers. But the smoke from fires can travel far distances, enveloping the populated clusters with polluted air.

In the 2019 regional survey, 25% of our California sample reported that they or someone they knew had experienced a wildfire event in the previous twelve months, which is not surprising considering that wildfires occurred in both northern and southern California. But slightly more than double that number (52%) experienced the smoke. This is nearly identical to the percentage in our survey who said they were exposed to smoke in other Pacific coastal (Oregon and Washington) and interior western states (Colorado, Wyoming, Montana, Utah, Idaho, and Nevada).

For many people the smoke is no more than an irritant, but for others the small particulate matter (PM2.5) can lead to serious health problems as it travels from an individual's lungs into the bloodstream. Studies have shown that exposure to PM2.5 can contribute to many bad health outcomes such as strokes, allergies, and lung and bladder cancer. Only half of the Californians surveyed in 2019 who experienced wildfire smoke claimed to have taken any precautionary steps when they were exposed to wildfire smoke. Most of those who claimed to take protective measures only took minimal steps such as paying attention to air quality reports or remaining indoors. But remaining indoors might not offer much protection if the windows are open or the condition of the home insulation is poor. The best protection would be donning an N95 or P100 mask or using an air purifier to clean indoor areas, but less than a third of the 52% who took any steps whatsoever adopted either of these measures. A few others claimed to have left town (18%) or consulted a health care provider (9%). These numbers suggest at the very least that there is much to be done to increase awareness and action among westerners with respect to taking protective actions.

In addition, there are important differences across race, gender, age, and socioeconomic status that must be accounted for when designing programs to protect people from wildfire smoke. Breaking down the 2019 California subsample of the regional survey further reveals that women, people over age sixty-five, whites, Asian Americans, the better-educated, and residents in the Bay Area were all more likely to have taken steps to protect themselves when experiencing wildfire smoke. Wealthier and better-educated people were more likely to have used N95 masks and air purifiers, while the poorer and less well educated relied on the far less effective cloth mask. Some of this is due to lack of knowledge, but N95 masks to some degree and air purifiers especially are costly items that can strain the budgets of people with limited resources.

A second obstacle to self-protection is partisanship. Some people do not want to believe the science about wildfire smoke because of their commitment to party or ideology. This can be seen more clearly by comparing the experience in California in 2019 to 2020. In 2020, a series of fires in the state sent plumes of smoke over a much wider area in 2020 than in 2019. As figure 3.1 shows, there were more residents who experienced both wildfire smoke and flames in 2020 than in 2019, which

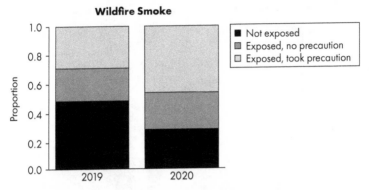

Figure 3.2. Proportion of respondents to the Bill Lane Center's 2019 western regional survey and 2020 California wildfire survey who had not been exposed to wildfire smoke, were exposed but took no precautions, or were exposed and took precautions.

corresponds with the objective data about the number, size, and duration of wildfires in those two years. Some patterns, however, were the same in both years. For instance, more people experienced the smoke than the flames in both years. Also, more people in 2020 when exposed to the smoke took precautions in 2020 than in 2019, but even so, a significant number still took no precautions at all despite the harmful effects of the smoke (figure 3.2).

Prior to the pandemic, there was a strong income and education trend in determining who wore a medical mask and who did not. But in 2020 that socioeconomic correlation diminished, quite possibly because of pandemic masking policies and the more widespread availability of cheap masks (figure 3.3). The partisan pattern is more disturbing. The gap between Democrats and Republicans diminished a bit between 2019 and 2020 but did not disappear, reflecting the general conservative Republican opposition to mask wearing in MAGA country.

The environmental justice problem, however, becomes more problematic as the cost of protection rises. Medical-grade masks are relatively cheap and are often distributed for free by employers, hospitals, and other institutions. Air purifiers, on the other hand, are quite expensive but are more effective against PM2.5 exposure especially indoors, where people are understandably reluctant to wear masks in their own homes.

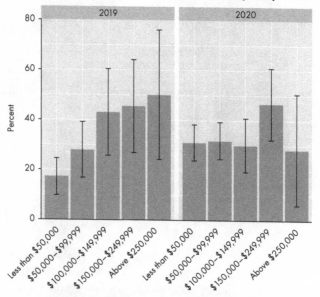

a. Took precaution: Wore medical-grade mask, by family income

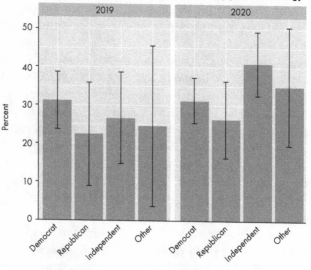

b. Took precaution: Wore medical-grade mask, by political ideology

Figures 3.3a and 3.3b. Percentage of Californian respondents to the Bill Lane Center's 2019 and 2020 surveys, disaggregated by income level and political ideology, who wore medical-grade masks to protect against wildfire smoke.

As figures 3.4a–c show, there were significant gaps in air purifier use by income, education, and type of home. The lowest-income groups were less able to afford air purifiers and more likely to live in badly insulated housing structures. Being less well educated makes it more likely that you are not as aware of the health consequences of breathing in PM2.5 and other particles in the smoke. And living in mobile homes or apartments makes it more likely that the smoke is accumulating in the interior of residences at high levels.

PARTISANSHIP AND WILDFIRE POLICY

There is substantial academic literature on the subject of whether experiences with extreme weather alter popular attitudes toward the existence, causes, and solutions to climate change.[18] The results of these studies are mixed to date, varying with the context and nature of the data used in each study. With respect to wildfires, we see that two key elements of risk do not coincide perfectly. The risk of losing life and property to wildfires is largely concentrated on residences and commercial properties that are located in or adjacent to wildland areas such as rangelands and forests. Hence, the perceived hazard of wildfire flames is high, but the exposure to risk is limited to a small segment of the electorate who were either burned out or at least forced to evacuate to save their lives. On the other hand, the toxic plumes of wildfire smoke from major fires or, worse, from fire clusters can spread widely across many areas. Wildfire smoke, given the broad population affected by it, might ultimately be a more important agent of attitudinal change than the threat of being burned out per se but only if people recognize the danger of exposure to wildfire smoke.

Earlier we noted that partisan divisions are so great that Republicans who are exposed to wildfire smoke are less likely to acknowledge it or take steps to protect themselves from its dangers. In addition, according to our data, Republicans are also generally less likely to favor government expenditures or regulation with respect to wildfires. This strong aversion to taxes and regulations carries over to using public funding to incentivize people to upgrade their homes or retreat to safer places, creating a partisan gap between Democrats and Republicans on these two policies.

Does the experience of wildfire flames or smoke in any way reduce this partisan gap? This work is reported in full elsewhere, but here we

a. Purifier usage by home type

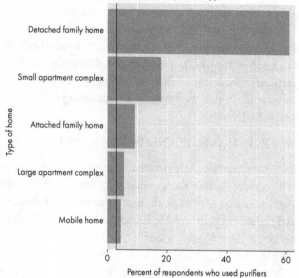

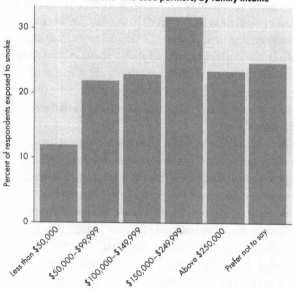

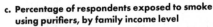

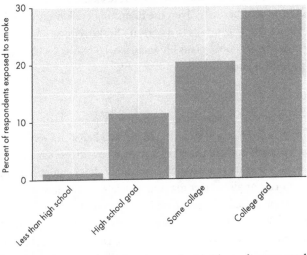

Figures 3.4a, 3.4b, and 3.4c. Percentage of respondents, disaggregated by home type, family income level, and education level, who used air purifiers to protect against wildfire smoke.

will consider only the final results as pertaining to respondents' actual proximity to fires (i.e., whether the respondent lives within ten miles of the fire) and the wildfire smoke intensity where respondents live to see if Republicans who had either experience were more likely to be in favor of using public subsidies as compared to those who had neither experience.[19] We find that proximity to the flames does alter Republican attitudes toward subsidies for home upgrades and retreat, but exposure to wildfire smoke does not.

What does this tell us about wildfire policy? The steps that jurisdictions can take fall roughly into three categories: education and voluntary actions, carrots (positive incentives), and sticks (regulatory compliance). Republicans who experience wildfires clearly have a greater understanding of the risks and are more interested in using public funds to enhance protections. Unfortunately, as noted several times, the number of such voters who fit into this category of risk is relatively limited. The danger of being burned out is relatively limited to those who live in or adjacent to wildland areas even if the hazard itself is high. Wildfire smoke by comparison has a wider exposure element but is lacking in the perception of hazard. Given the mounting evidence that wildfire smoke is actually hazardous to a person's health, educating the public about this danger could plausibly increase the public's willingness to fund protective policies.

———

Several important implications are evident from people's attitudes toward wildfire policy. To begin with, the widespread belief in California that people should be allowed to make their location and resilience decisions regardless of wildfire risk is highly problematic. It is compounded by the fact Californians also prefer making people safer in place rather than moving them to safe places. These individually chosen risks result in higher socialized costs, such as the expenses and dangers of firefighting, the insurance losses that drive up rates, the care for displaced populations, the liability costs due to downed utility power lines sparking fires, and so on. Experiencing a nearby wildfire at least inclines some Republicans to be more supportive of using public funds to upgrade their properties to be resilient in place (e.g., home hardening, defensible space, and son) and to enable people who want to leave and rebuild elsewhere to do so. But the number of residents who live next to or within WUI areas is not sufficiently big to have a large impact on public opinion.

Of course, if the costs of fighting fires and paying for their cleanup continue to grow, the socialized costs of individual freedoms will become more apparent over time and could change the prevailing perspective of the right to live where you want. Wildfire smoke is in many ways comparable in danger to secondhand smoke from tobacco and could potentially have a more immediate political impact at a scale that would change more people's minds about the seriousness of the drought and wildfire problems associated with global warming. However, the evidence indicates that so far exposure to wildfire smoke has not significantly impacted the willingness to use public funds for wildfire resilience measures and in some cases even diminished support for public funding of certain resilience measures. And the partisan gap does not change. Republicans were less likely to agree that the smoke is harmful or that people need to invest money to protect themselves from that smoke.

What, then, will it take for personal wildfire experiences to produce greater partisan agreement on wildfire adaptation policies? Clearly, the continued expansion into WUI areas and the effects of a warming climate will increase the number of evacuations and personal experiences with wildfires. Perhaps the expense of fighting these fires will alter perceptions about requiring and publicly subsidizing stricter preventive measures. Or perhaps the wildfire concerns of those who live in WUI areas will be bundled with those who suffer from more extreme heat, water scarcity, or sea level rise to form an extreme weather coalition that logrolls each other's concerns into a winning coalition in support of extreme weather resilience.

Wildfire smoke has the reach but not the perceived hazard level, particularly among Republicans. This perception might be changed with better information outreach, but the problem of motivated reasoning will likely undercut such efforts. As with the rising costs of fighting and recovering from wildfires, the expense of dealing with the health costs associated with wildfire smoke may eventually increase support for wildfire adaptation measures in the future. In the meantime, the partisan gap on climate change resilience will likely persist.

CHAPTER 4

RELUCTANT RESILIENCE

Coastal Threats and Delayed Solutions

Residents of the arid and semiarid West must also cope with extreme flooding in addition to severe periodic droughts. In the Bill Lane Center 2019 regional study, 11% of westerners claimed to have personally experienced extreme flooding during the previous twelve months, and 25% believed that it had happened somewhere in their state. Heavy rains can lead to flash flooding in low-lying areas in both inland and coastal regions, sending dangerous torrents of water down previously dried-up creeks and rivers.

Too much water is as much a fact of life in the American West as is too little water. Western weather is marked by many contrasts over time and space. Precipitation beyond the 100th meridian varies widely by season (i.e., winter versus summer months), year (i.e., wet versus dry years), and microclimate (i.e., the Northwest versus the Southwest), which has made western water management complex and technically challenging. Climate change adds more uncertainty to the task.

Coastal states (i.e., California, Oregon, and Washington) face the duel challenge of future sea level rise and more extreme storms. This is comparable to what Atlantic and Gulf Coast states face but is different in one respect that has political consequences. Storms along the Pacific coast are seldom if ever hurricanes. Rather, they take the form of a series of storms carried along by atmospheric rivers that deluge the land for days on end with intermittent heavy rains. These storms generally inflict moderate levels of property damage. Absent violent hurricane winds, they are not as life-threatening and dramatic as the East Coast hurricanes. The political consequence is widespread complacency and mostly resilience procrastination in Democratic strongholds that are ostensibly committed to dealing with climate change in a serious way.

Future sea level rise along with enhanced storminess will worsen flooding along the Pacific coast, likely causing extensive erosion and damaging critical infrastructure and property. The costs of policy inaction are potentially substantial for several reasons. The Pacific coastline from the Mexican border to the Canadian border is vast (i.e., 1,293 miles, or 7,863 miles if one includes tidal inlets) and densely populated around major urban areas. The coastline region also contains a great deal of critical infrastructure such as power plants, roads, bridges, and wastewater treatment facilities. The potential economic costs from sea level rise along the Pacific coast could amount to up to $1 trillion by 2100 if the right protective measures are not taken soon enough.[1]

Climate change politics plays out differently in the coastal West than in the interior parts of the region. Coastal communities in the West are more Democratic, liberal, and affluent as compared to the inland interior regions. Partisan differences between Republicans and Democrats are still evident in the Pacific West but matter less because Democrats outnumber Republicans by a considerable degree, especially in cities such as Los Angeles, San Francisco, Portland, and Seattle. Voters in these areas mostly believe that climate change is real and that human activity is responsible for it to a significant degree. But while there is much talking the talk about the need for resilience measures under more extreme weather conditions, there is not as much walking the walk as one might think.

In part this is because the environmental community initially put more emphasis on addressing the causes of global warming than adapting to it for fear that doing otherwise would give people the wrong idea about living with global warming rather than mitigating it. Adaptation would be a fool's errand if you only adjusted to the problem at hand and ignored the potentially controllable reason (i.e., carbon emissions) for why the weather is getting more extreme. That thinking has changed, and there is now widespread recognition in environmental circles that adaptation is critically important. But even as the realities of extreme weather have kicked in, local communities have been slow to take the necessary measures to protect against the compounding threats of high sea levels and enhanced storminess. The reason is not so much partisan denial as it is a concatenation of routine political failings: that is, a lack of urgency about future problems, conflicting priorities within the community, agency processes that too easy enable obstruction, and jurisdictional parochialism that inhibits effective cooperation between communities that share a common problem.

THE COMPOUNDING IMPACT OF STORMS AND SEA LEVEL RISE

Despite progress in scientific forecasting of extreme weather, predictions about the timing and intensity of future extreme weather events are necessarily subject to numerous contingencies and uncertainties. The most critical one is the progress the world makes in reducing greenhouse gas emissions, as that will impact the rate and intensity of storms and sea level rise. While nations have pledged ambitious goals once again at the United Nations Climate Change Conference in Glasgow (COP26) meetings, the record of accomplishment to date is not encouraging.[2] Many scientific models build this uncertainty into their predictions by considering a range of possible scenarios from "business as usual" (i.e., negligible improvements from the status quo) to the most optimistic ones that assume that the world holds global warming to under 1.5 degrees Centigrade. In the language of politics, the level of greenhouse gas emission mitigation is a "known unknown" with respect to planning: we know that it is critical to planning protective infrastructure, but we do not know what progress will be made in the coming years.

There are other connected unknowns as well, such as the melting of icebergs. Warming seas are causing massive glaciers at the South Pole to melt, transferring water from land to the oceans. In some scenarios, this could cause several feet of sea level rise. While the melting rates in the Arctic and Antarctic regions are subject to a great deal of scientific research at the moment, it is hard for scientists to make a reliable forecast as to the rate and amount of melting.[3]

Another type of uncertainty that will affect flooding along the coastline is the rising and sinking of land due to such factors as plate tectonics, proximity to large Alaska glaciers, and groundwater depletion. Above Cape Mendocino along the northern coast, the subduction of the ocean plate under the northern plate causes the land to rise at a rate between 1.5mm and 3mm per year.[4] This could offset impact of sea level rise in those areas to some degree.

However, below Cape Mendocino the plates are sliding past one another, causing the land to sink at 1mm per year, which would worsen the flooding problems in already low-lying areas. A major earthquake along the fault lines there could cause the land to drop and the sea level to rise by several feet. Human activity affects land levels as well. The land is subsiding in some coastal areas due to substantial groundwater withdrawal for farming and residential water supply purposes. This not only increases the flooding but also escalates the rate of saltwater intrusion into coastal aquifers.

In conjunction with sea level rise, there is also the threat of more serious storms coming from the warming Pacific waters. El Niño storms already cause extensive flooding in low-lying areas due to larger waves, higher tides, and bigger storm surges. Warming oceans will likely increase the frequency and intensity of these storms, but by how much is yet another element of unknown risk. Large El Niño events already raise coastal waters by four to twelve inches during the winter months, and it will likely be much greater in the future. Then again, it is also possible that the Pacific storm track could move north, sparing California and worsening the situation for Oregon and Washington.

From the point of view of a public official, these various "known unknowns" greatly complicate the task of planning and advocating for expensive infrastructure. Scientific projections are useful exercises, but leaders of communities must choose one scenario as the baseline when they make project decisions. If the projected sea level rise in a given area is in the range of three to six feet, does the community plan the height of its levee for the low or high end of the distribution? Should the plan include an extra margin of height for safety? Do you choose the 2030, 2050, or 2100 sea level rise scenario? Select a high-end projection, and there is the political danger of incurring unnecessary costs. Rely instead on the low-end projections, and the official could be vilified for failing to protect the community adequately.

Without budget constraints and short-term electoral pressures, the safest policy solution to sea level rise would be to act quickly, assume the worst, and build redundant levels of protection. But local officials do not operate in such an ideal world. Instead, they must deal with electorates replete with antitax activists, budget hawks, and well-meaning wildlife advocates who will object to any and all measures that potentially harm wildlife and the natural ecosystem.

Looming over all elected officials is a time perspective dictated by short terms in office and the threat of a recall election if an official charges voters with a tax for something they do not want at the moment. Most of the voters who will determine that official's fate in the next election won't live long enough to reap the maximum protective benefit in 2100. Moreover, those benefits are counterfactual in nature: they prevent bad things from happening as opposed to delivering immediate tangible benefits such as a tax cut or new playground. Counterfactual benefits are harder to prove to skeptics. There is no proverbial bird in hand. It remains in the bush.

Nuisance floods (i.e., less than catastrophic events) are common in low-lying coastal areas and in theory should be the wake-up call the public needs. But these tepid natural disasters at best crack open a narrow, impermanent policy window. Sometimes the best self-preserving political strategy is to hold meetings but take no action. This is what currently happens in many coastal communities.

THE DENSITY OF PEOPLE AND INFRASTRUCTURE IN COASTAL COUNTIES

There are several aspects of the coastal resilience problem that set it apart from the drought and wildfire challenges discussed earlier. The first is that coastal flooding, as compared to the wildfire threat, impacts heavily populated areas to a greater degree. This means that the sheer number of people and total property value at risk is more substantial along the coast than in inland areas of the Pacific states. Seventy-five percent of Californians, for instance, live in coastal counties despite recent growth in the interior driven by the lower costs and wider availability of housing there. There is no escaping climate risk in California. Moving to the interior means dealing with more extreme heat and wildfire smoke that accumulates in the valleys. Live on the coast, and it means dealing with potential earthquakes and floods due to sea level rise.

A central driver of locational risk is the amenity allure of living near scenic wildland or ocean water. According to evidence collected in a 2019

California Legislative Analyst's Office report, at least 480,000 people live in low-lying coastal areas that are at risk of inundation by a one hundred–year coastal storm event, especially in southern California and around San Francisco Bay. The odds of such an occurrence could increase to as much as one in five years by 2050.[5]

While many of these at-risk coastal communities are relatively affluent (e.g., Imperial Beach, Coronado, Mission Beach, Venice, Oxnard, and San Francisco), this is not the case in all instances. Living inland is generally cheaper in the three Pacific coast states, but there are significant low-income and minority concentrations living near ports, refineries, and industrial zones. A 2012 study estimates that around San Francisco Bay, the risk to disadvantaged communities included "more than 9,000 renter-occupied households, over 2,500 linguistically isolated households, over 2,000 households with no vehicle, and over 15,500 individuals living in households earning less than 200 percent of the federal poverty level."[6]

A related aspect of the high population concentration in coastal areas is the density of critical roads, power plants, wastewater treatment plants, schools, and hospitals. Knocking out critical infrastructure can adversely affect the lives of many people who live and work outside the flooded and damaged areas. Major airports in Oakland, San Francisco, and San Diego, for instance, are quite vulnerable. A sixteen-inch sea level rise at high tide could inundate the Bay Area airports, halting commerce and tourism for potentially weeks at a time. An ocean economy valued in excess of $40 million would be at risk. There are 114 state parks covering 340 miles of the coastline. The erosion caused by sea level rise would threaten state beaches that currently generate more than $3 billion in revenue. Overall, estimates of possible economic losses for California tourism are at least $14 billion.[7]

Erosion of the coastal area also equates to permanent land loss, precluding the option of rebuilding structures in the same location as is often done in the case of inland wildfires. While some might argue that this would be for the best, widespread sea level rise flooding could cause high property and sale tax losses for some coastal communities if people and businesses either could not or did not want to rebuild in the same local jurisdiction.

Clearly, there is an urgency to incorporating expected climate change impacts into local community planning more effectively. The California state government has for over a decade issued various public reports outlining the dangers of sea level rise and increased storminess. Since 2014, it has provided four major streams of coastal protection funding totaling $67 million for projects, grants, and more stringent regulatory review. Despite those efforts, the Legislative Analyst's Office found that most coastal communities were still only in the earliest stages of adaptive planning.[8]

Even fewer coastal communities had actually implemented any measures. Despite the exposure to potentially widespread property damage and economic loss, California local communities have for various reasons found it difficult to devote the time, effort, and resources necessary to develop plans and measures to protect themselves from sea level rise flooding.

But if the state's voters recognize the sea level challenge, why doesn't this translate into quicker action at the local level? A core reason is that local governments retain the primary authority over land use in their territories. The federal government and state governments have some regulatory authority due to environmental laws and the infrastructure that they own and control. But local governments jealously guard their control over residential and commercial development because this generates the property and sales tax that they depend on to fund local budgets. Adaptive steps such as strategic retreat and zoning out new development in likely flood areas run counter to the community's economic interests, much like building new housing in high-risk wildland areas. Without addressing the perversities of property tax policies that create huge gaps in assessed value between older and new properties, local communities will regard new housing and commercial property as an easier path to revenue enhancement as opposed to levying special taxes on existing properties.

Fiscal politics complicate resilience matters in other ways as well. While various studies have shown that taking protective measures before natural disasters occur would spare local communities even bigger losses down the road, elected officials continue to find it difficult to justify to taxpayers why they need to pay for hypothetical future protection when there are other more immediate pressing police, fire, and social service needs. In some ideal world, elected officials would look beyond what voters want at the moment to what they need in the long run, but that is not the world we operate in at the moment.

Even assuming that a local community wanted to be proactive with respect to sea level rise, there are other obstacles to overcome. It would be better to work cooperatively with other neighboring communities along the coast to ensure fairness in the protection being offered. A coastal community that unilaterally puts a seawall up can protect its residents and commercial areas quite well, but if neighboring communities do not also act, they can experience worse flooding as a consequence. In the wildfire case, no protective measure can prevent hot flying embers from crossing over jurisdiction lines and causing extensive damage to property and people. But water can more effectively be deflected away from one area into another by building seawalls or levees. This means that there is a strong incentive for coastal communities with the motivation and sufficient capacity to act unilaterally to adopt protective measures regardless of the impact on neighboring communities.

This problem is compounded by the advantages that wealthier cities have in raising the funds for expensive coastal armoring by issuing bonds or levying special property taxes. Wealthier communities also have enough city staff to look out and apply for matching grants offered by the federal government and state governments. Such communities also are more likely to have competent in-house engineering capacity or the resources to hire private consultants, which means they are also more likely to win grant competitions that are judged on quality technical design.

State and federal authorities can and sometimes do try to counter these problems. The California legislature and the Department of Water Resources recognized that the Integrated Regional Water Management program favored high-income and predominantly white communities after conducting early rounds of a matching grant competition to develop the state's freshwater resources. They responded by altering the grant scoring in order to favor more inclusive and collaborative water grants between wealthy and disadvantaged communities. The Integrated Regional Water Management program example demonstrates that the federal government and state governments can use grant design and subsidies to offset the go-it-alone incentives for wealthier communities and incentivize more equitable collaboration if they make that a priority as the state legislature did.[9]

THE THICKNESS OF REGULATORY REVIEW
IN COASTAL COUNTIES

Aside from information about sea level rise that governments provide through published reports and the nudge that offering grants provide, governments at all levels have regulatory powers they can wield due to laws such as the National Environmental Policy Act, the California Environmental Quality Act, the and Clean Water Act. These laws all serve important purposes. However, it is also increasingly apparent that the jurisdictional complexity of getting permit approvals from numerous local, state, and federal entities can be used to delay or even obstruct the deployment of much-needed resilience infrastructure.

A good illustration of this is the case of the San Francisquito Creek that runs adjacent to the Stanford University campus. In 1998, a forty-five–year flood caused by a series of winter storms damaged 1,700 properties in Palo Alto, Menlo Park, and East Palo Alto. The creek had overflowed seven times since 1910, so the communities came to realize that this was a recurring problem that could not be ignored, especially given the prospects of even larger flooding events in the future. The US Army Corps of Engineers estimated that a one hundred–year flood could cost twenty-five times more than the $28 million 1998 event. And none of this accounted for projections about future sea level rise in the area.

The Army Corps of Engineers recommended that levee walls surrounding the creek be redesigned to prevent future flooding. Five local entities—two special water districts plus Menlo Park, East Palo Alto, and Palo Alto—joined together to form a joint powers authority (JPA) to deal with this problem. These kinds of arrangements allow communities to formally commit resources to and establish a joint governance structure for handling problems that spill across local boundaries. The JPA eventually formulated a plan to protect against the one hundred–year flood by widening the flood channel and rebuilding the levee. But the plan had two obstacles. First, it meant taking away some of the land on the Palo Alto side from its municipal golf course. To appease the golfers, the city had to pledge additional money to enhance the course to compensate for the land loss.

The second problem was that the adjacent wetland on the East Palo Alto side of the flood creek contained two endangered species: the clapper rail and the salt marsh mouse. Any design had to ensure that these species would not be hurt. In order to get the permits to undertake this work, the JPA would first have to undergo a review under California's California Environmental Quality Act and then a second round of four separate federal and regional permits or biological opinions from the Regional Water Quality Control Board, the San Francisco Bay Conservation and Development Commission, the California Department of Fish and Wildlife, and the US Army Corps of Engineers (in consultation with the US Fish and Wildlife Service and the National Marine Fisheries Service). Each agency had to be consulted separately, and in some cases accommodations for one agency would require going back to another agency to clear any subsequent changes with them.

All of this would be complex enough, but modern permitting is also conducted with a great deal of transparency and opportunities for public comment. Various nonprofits and NIMBY and other neighborhood groups used the many steps in the permitting process to relitigate points they made in earlier stages of the process. Some of these objections were clearly disingenuous attempts to gain bargaining leverage with respect to other goals such as marshland restoration or dam removal. For instance, opponents to the JPA plan raised the question of why the levee wall on the side of the more disadvantaged East Palo Alto was several inches lower than on the Palo Alto side.[10] The JPA's experts explained that this was because the levee wall on the Palo Alto side would be new and hence expected to settle a few inches, whereas the one on the East Palo Alto side was largely constructed from the existing wall. Nonetheless, the same pretextual objection was raised in subsequent stages of the review by predominantly white environmental groups, not by the residents or officials of the minority communities in East Palo Alto.

From August 2012 to December 2014, the JPA repeatedly went back and forth with agency officials answering objections and concerns raised by community opponents. In the meantime, the project costs were increasing, and properties remained at risk due to the effort of some to use this issue to gain the upper hand on other environmental disputes.[11] Many people were frustrated that a project that would protect lives and property could be delayed in some many ways.[12] The project was finally completed in December 2018, twenty years after the forty-five-year flood and six years after an intervening flood in 2012 that required disaster assistance.

Permitting complexity is partly rooted in the high degree of specialization in modern agency regulation. Multiple agencies at various levels of government often have jurisdiction over different pieces of the environmental permitting process. This creates higher transaction costs, widely divergent agency perspectives, and lengthy delays. Permitting can be expedited by convening agencies to meet and make decisions together. This prevents the ping-ponging of decisions from one agency to another that occurs when agencies make separate decisions sequentially as they typically do on large projects that involve multiple permits. Another possible fix is to agree to and set out safe harbor guidelines that allow projects to go forward without repeating lengthy and expensive biological reviews as long as they conform to previously established parameters. And yet another is that under emergency conditions, normal state regulatory review can be suspended completely, as recent California governors have done during major droughts. But emergency powers are usually contingent and time-bound to the context of a crisis and hence do not provide stable solutions for chronic problems.

The problem of regulatory fracture is exacerbated in wealthier and highly educated urban areas by the ability and willingness of citizens and some neighborhood groups in these areas to use environmental obstruction for political purposes. It is well established in political science literature that highly educated, older, and affluent citizens are more likely to participate in politics consistently than are younger, poorer, and less well-educated citizens.[13] People are more likely to be active when they have the time and resources to spare and the white-collar interpersonal people and organizational skills to be effective.

Efforts to open up administrative procedures in the interests of making them fairer and more transparent have unintentionally created a situation in which interest groups and others can thrive. Monitoring technical procedures such as environmental permits is not for the average citizen. The details are complicated, and the time and effort required to do this effectively is usually substantial. By comparison, well-resourced nonprofits, neighborhood groups, and businesses with a stake in the outcome are usually very attentive to public proceedings and ready to mobilize against what they do not like. This creates a selection bias in the information and feedback that agencies receive when they invite public input.[14] Over time, interested parties with material interests at stake become more adept in manipulating the permitting process in order to delay and obstruct outcomes that they oppose and advocate for those they favor. And as the San Francisquito Creek story illustrates, when projects require both state and federal permits, interested parties that fail to get what they wanted at one level can try again at another.

This kind of problem also plays out in rural areas with respect to water infrastructure (e.g., with dams and reservoirs) and wildfire vegetation management (e.g., neighborhood objections to smoke from prescriptive burns), but the thickness of the stakeholder environment varies greatly with context. In highly educated liberal communities, the nonprofits and NIMBY neighborhood organizations abound, presenting a surprisingly tougher resistance to climate resilience projects than one might imagine given the political bent of those areas.

Communities in the Bay Area of California are fertile ground for NIMBY resistance, especially when residents fear the possibility that levees or other gray infrastructure might adversely affect the quality of life in their neighborhoods or the value of their homes. One recent of review of NIMBY obstruction in California found that half of the lawsuits under the California Environmental Quality Act targeted public rather than private projects, including transit and infill development projects that were meant to encourage mass transportation and discourage lengthy daily work commutes by car from outlying counties.[15] Given that transportation is a major source of emissions in California and that the opposition to them often emanates from liberal communities' belief that climate change is real, this behavior seems at least inconsistent and at worst hypocritical.

In most cases the objections do not emanate from state and national environmental advocacy groups, as they are usually publicly committed to reducing transportation emissions. More commonly, obstruction comes from neighborhood groups that invoke the language of environmental protection to add legitimacy to their efforts to halt infill projects through lawsuits and obstructive tactics. The review, quoting the views of one land expert, notes that

NIMBYs comprised by far the largest number of project opponents, particularly for infill projects. NIMBY opponents were often characterized as "older" or "wealthier" or "less ethnically diverse" than the part of the population that would benefit from the challenged project, particularly for urban schools, parks, and multifamily housing projects. As a noted land use expert has observed, "[t]he people who are most apt to fight things have six-figure incomes and nice houses and college and post-college degrees." NIMBYs and their advocates are often personally impassioned about protecting "their" environment, defining the "environment" as their local community.[16]

National and state nonprofit groups with perfectly legitimate environmental concerns—such as wildlife issues around solar and wind projects, the impact of desalination projects on ocean biota, and displaced beach erosion from proposed seawalls—also use similar tactics to delay utility-scale solar and wind projects when wildlife issues need to be considered. Whatever the merits of these objections, the vulnerability of permitting processes to obstruction will be an important problem for years to come as the urgency of meeting both resilience and decarbonization goals comes into sharper focus.

Current permitting processes under the National Environmental Policy Act and similar state legislation serve critical purposes by soliciting public input and requiring transparency about public projects. The permitting processes force public officials to consider seriously ways that public projects can minimize adverse environmental consequences. The threat of losing lawsuits or even paying the expense to win them provides additional incentives to design public infrastructure projects carefully. But these benefits have to be weighed against the consequences of delaying the implementation of critical resilience measures, especially in thick nonprofit network settings such as in Pacific coastal counties.

RESILIENCE POLITICS IN LIBERAL SETTINGS

Coastal counties in the West are blue in political orientation. In California, Oregon, and Washington, the divisions between highly Democratic and Republican areas fall along an east-west, coastal-inland axis. This is partly a legacy of the way far western state economies evolved due to the importance of their harbors for Pacific Rim trade and military defense in the latter half of the twentieth century.[17] The West initially developed around agriculture, timber, and mining industries. During World War II, federal money poured into far western states in order to defend the Pacific front from Japan and later to support the defense industry during the Cold War. In recent decades, the growth of high tech in San Francisco, Seattle, Portland, Los Angeles, and San Diego further widened the divergent political orientations in coastal and inland areas in California, Oregon, and Washington.

These economic forces also attracted domestic and international migrants, which led to increasing racial and ethnic diversity. African Americans, for instance, were drawn from the South into urban California areas such as Oakland, Los Angeles, and Richmond, where they were needed for the war effort. Latino farmworkers ended up in the Central Valley and other rural areas of the far West thanks to the Bracero Program.[18] But subsequent waves of documented and undocumented Latinos migrated into urban manufacturing and service sectors, especially in southern California. Over time this heightened racial and ethnic diversity and solidified Democratic control in cities, while the interior rural areas of these states absorbed white flight from urban areas and became more Republican over time. In effect, these trends mirrored the national patterns of party and geographic sorting: the party coalition realignment of southern, rural, and Reagan Democratic white voters into the Republican Party and the homophilic spatial clustering by party, ideology, and race.[19]

The blueness of the coastal counties has several implications for resilience politics along the coast. The preponderance of Democrats in these areas creates a wider acceptance of the scientific projections about future sea level rise as compared to the higher rates of climate denial in the more conservative rural white Republican interior. This in turn translates into more voters understanding the threat of coastal flooding, which should in theory pave the way for more actual policy implementation, and the solutions they prefer are divided along party lines.

When respondents in a 2019 Bill Lane Center California sea level rise survey were asked about whether they believed they would be personally harmed by sea level rise, 66% of Democrats versus 40% of Republicans in California indicated that they thought they would be harmed either a great deal or a moderate amount. So, partisanship matters with respect to these issues, but it is not quite the stark divide that national party rhetoric often suggests. A statistical model that controlled for different respondent characteristics reveals that older and less well-educated Republicans are more likely to be skeptical of sea level rise harms than younger and better-educated ones.[20] These skeptical Republicans are found in greater abundance in the eastern inland areas of the West Coast, while Republicans who live in urban and coastal areas tend to be more receptive to the message about future sea level rise. In other words, between the high concentration of Democrats and urban Independents plus more moderate Republicans in the coastal counties, there is a fairly broad cross-party understanding about the danger of future sea level rise.

Concern about sea level harm is not correlated with how close you are to the water. There was no relationship between distance from the shoreline and the fear of personal harm from sea rise flooding.[21] This is perhaps because earthquakes and previous floods have educated California residents and the press about the widespread regional consequences of natural disaster damage to roads, airports, and power plants.

Partisan differences are somewhat greater when it comes to questions of what to protect and how to pay for it. The survey asked "how important do you think it is for the government to invest public funds in the following strategies?" and provided the options of reinforcing bridges, highways, and roads; fortifying wetlands; protecting natural sensitive habitats; upgrading water and wastewater treatment plants; protecting private homes; protecting commercial properties; and moving or rebuilding airport runways. Across the whole sample, Californians assigned the highest important to highways, bridges, and roads (62% very important) and wastewater facilities (60%) and the lowest to protection to private homes (29%) and commercial properties (19%). Less than a quarter of them said that spending public money for these purposes was not at all important, which suggests little vehement opposition to spending public money for protection from coastal flooding.[22]

Republicans, however, were significantly less supportive than Democrats of using public money in most instances, as in the case of wildfires. The gap between Republicans and Democrats was smallest in regard to using public funds to protect homes and commercial properties and largest in regard to protecting wetlands and natural habitat. The lower partisan differences about protecting homes and businesses from sea level flooding reflects the fact that many Democrats, even liberals, share

Figure 4.1. Percentage of respondents, disaggregated by political ideology, who were willing to take certain preventative steps against natural disasters.

the prevailing belief that people who live in risky areas are responsible for their residential losses (figure 4.1). Not surprisingly, there is sympathy for relief efforts after natural disasters, rewarding politicians for making Federal Emergency Management Agency relief available. On the other hand, natural capital solutions (i.e., using public funds for nature-based protective measures such as enhancing wetlands in at-risk areas) evoke more partisan responses, which comports with the large general partisan divide over conservation and wildlife protection.

What happens when Republicans fear the consequences of sea level rise flooding? Personal experience makes partisan gaps between Republicans and Democrats become statistically insignificant in all cases except protecting sensitive natural habitat. In short, Republicans would join Democrats in supporting public action to enhance flooding protection but not for all measures or by all revenue avenues. The gaps between

Democrats and Republicans remain significantly large with respect to using public funds to help property owners obtain insurance, buy out their endangered properties, or facilitate their relocation to safer places. This is even when the public funds would be targeted for assisting low-income households (unlike in the wildfire case).

Similarly, there is a large gap over how to pay for protective measures, including all forms of sales and parcel taxes, without any difference between those who feared versus those did not fear the harmful effects of sea level rise. The one glimmer of hope in this fiscal picture is bond measures. The partisan gap is smaller to begin with and becomes insignificant for both Republicans and Independents when factoring in personal fear of harm from rising sea levels. Among all sample voters, issuing municipal bonds was the only financing measure that achieved majority support. This very much reflects a decades-long modern trend in California public finance to put infrastructure projects into bond measures on the ballot to be decided by the public. Many of these measures have been approved by California voters over the years, replacing the state's general funds for parks, roads, schools, prisons, and other infrastructure. In essence, the California public prefers taking loans to paying out of pocket for these necessities.

Other policies, while controversial, are not so divided along partisan lines. One example is whether to offer gray or green solutions. Gray options include levees, dikes, seawalls, revetments, storm-surge barriers, drainage systems, and pump systems, while green options consist of beach nourishment, enhanced wetlands, and horizontal levees. The attraction of green solutions is that they can be designed to offer recreational and ecological cobenefits that make protective infrastructure projects more attractive to the taxpayers who have to pay for them, potentially neutralizing NIMBY concerns about lowering the quality of life and property values in their neighborhoods.

In heavily developed areas and around critical infrastructure, however, the gray solutions such as levees and seawalls are more feasible and have a proven track record of reliability, whereas the green solutions generally do not. Given the size of the potential economic losses at stake in many coastal counties and the competition for city and county funds from other pressing policies, public officials are less likely to take a chance on novel natural capital solutions.

And no matter what public officials do, both gray and green solutions will engender many forms of political opposition. This is the iron law of public projects. When New York City tried to move forward with a $1.5 billion project to protect the Lower East Side of Manhattan by elevating a park and adding flood breaks and seawalls, a group of local residents and activists sued the city, claiming that the changes would "destroy the park" and result in its partial closure over a five-year construction period.[23]

One last important lesson from public opinion data is that as with wildfires, it is more popular to restrict future growth in risky flood zones than to remove existing property and move it to a safer place. Overall, Californians favored restricting new residential (66%) and commercial (65%) development in high-risk flooding areas. They were also opposed but a little more sympathetic to allowing people to rebuild in areas that had flooded (42% in favor of forbidding rebuilding vs. 34% opposed). But they were much more reluctant to force the hand of those currently living in areas that are at risk of damage due to future sea level rise. Californians also opposed mandating flood insurance (48% against, 36% for) and forcing those who live in flood zones to retreat and relocate (60% against, 24% for). All of this suggests that the easiest political path is to limit new development, not existing properties, but that way forward leaves a lot of exposure to flooding risk on the table.[24]

Stepping back for the moment, we see that some of the puzzle about why Californians can acknowledge the problem of sea level rise but not act as quickly and effectively as they need to is the narrow path that public opinion has created for local community leaders. Limits on future development entail giving up on much-needed revenue for local community budgets. And there are revenue constraints as well. It's okay to use public money to build levees and enhance marshes, but don't raise taxes; put it on the credit card (i.e., bond measures), please. Some communities manage to find their way through this narrow path, but others do not. What happens then?

COLLECTIVE SOLUTIONS AND
FIRST-MOVER PROBLEMS

While public recognition that the sea level rise threat is relatively high in coastal communities, various impediments have slowed local community responses but not in all places. Some communities have moved ahead with planning due to increasingly dire circumstance, while many others have not. These concerns about inaction are echoed in scholarly literature as well. A study about local community adoption of climate action plans—state-encouraged "comprehensive roadmaps" that outline specific actions that public entities plan to take to reduce emissions—found that twenty-four of fifty-eight counties had still not devised climate action plans a decade later.[25] Two factors seemed to explain support: larger cities with more capacity and the percent of Democrats in voter registrations. Moving from planning to the implementation of projects requires community motivation and government capacity: convincing voters that there is an urgent need, finding the money to pay for the project, deciding the type and placement of coastal protection, getting the permits from numerous agencies, and contracting with construction firms to get the job done.

The cities that make their way through this maze of tasks have the necessary level of both motivation and capacity, while those that are lacking in one or the other areas are less likely to move ahead. Given what we know about variations in the political orientation and capacities of local communities, the more wealthy and liberal communities are more likely to be the first movers.

In some environmental situations, first movers are the "chumps"; they pay the cost of trying to clean up or solve an environmental problem but cannot exclude others from enjoying the benefits of those actions. This is roughly the situation that California finds itself in with respect to emission reductions. The state has invested heavily in renewables and reduced its emissions by 10.5 metric tons person, a 25% emissions reduction since 2001. Other states that have not tried as assiduously to reduce their emissions still benefit from California's efforts.

There are other instances, however, such as armoring shorelines in which first movers get an exclusive benefit from their actions while creating negative consequences for other communities. Seawalls and levees can protect the residents of a given community but at the cost of displacing the water to other places along the shore, thereby worsening the neighbor community's flooding level and related property damage. Therefore, it would be preferable to coordinate the actions of all who might be impacted in order to avoid situations such as these. But sometimes there are only bad choices.

Consider the case of Foster City. Constructed with thirteen million cubic yards of compacted sand around an existing island in San Francisco Bay, Foster City is a relatively affluent predominantly white and Asian community that lies just south of the San Mateo Bridge.[26] In 2014, the Federal Emergency Management Agency revised its flood maps and deemed that the existing thirteen-foot levee would have to be increased in height or else the agency would require residents to purchase flood insurance that would cost an additional $2,000 to $3,000 above what residents were already paying. Faced with this prospect, the city put a $90 million measure on the ballot to raise the levee by an additional eight feet. Residents would pay off the bond with a property tax assessment of $40 per $1,000 of home value.

The city acted unilaterally to put this on the ballot despite concerns from experts and city engineers around the Bay Area. Building the wall, critics complained, would displace twenty thousand acre-feet of water onto other communities around the area. Moreover, it was not clear that the levee would actually provide adequate protection, since the city based its levee design on a lower-end estimate of likely Bay Area sea level rise and did not account for additional flooding from increased groundwater saturation or for water coming from the slough on the highway side of the city.

But residents of Foster City were faced with two imperfect choices: move forward with constructing a levee to avoid the Federal Emergency Management Agency's insurance rate hike despite potential adverse effects on other cities or pay the new insurance rates while waiting for some unspecified amount of time for a multidistrict solution. Given the fiscal consequences of delay and the uncertainty of collaboration, it was not surprising that Foster City residents voted to act unilaterally.

The crux of many climate change resilience problems is that local governments have the power to make land-use decisions, and any effort to make coordinated decisions would require some entity to be in charge. However, there is no overriding entity in charge and able to coordinate the actions of the many local communities around the Bay Area. There is, as Marc Lubell has coined it, a "governance gap."[27] He identifies a network of 103 projects with 512 unique actors working on sea level rise in the Bay Area. The consequence is lots of planning and study but no coordinated decision making.

In theory, a state agency or the governor and the legislature might be able to do the coordinating job, but cities are loath to surrender their sovereign control over land use to the state. Local government entities that provide expertise and advice to member cities such as the Association of Bay Area governments have neither the money nor the authority to wield power and influence in this area. The regulatory authorities for enforcing environmental law—the Bay Conservation and Development Commission, the regional water control board, and the US Army Corps of Engineers—are too feared by local authorities and fractured among themselves to be given the lead role. And while the Bay Area universities and nonprofit stakeholders convene regular meetings on the topic, they are also not able to make decisions.

In sum, there is no shortage of effort and concern regarding sea level rise in California, but the actual implementation of projects has developed at a snail's pace. When the situation becomes dire enough for an individual city with sufficient capacity and resources, local communities find solutions for themselves, as Foster City did. But unilateral solutions increase the inequity across communities, as the wealthier ones will be more capable of taking unilateral action than the more disadvantaged ones (e.g., East Palo Alto and Richmond). The levees and seawalls that wealthier communities can construct can leave poorer neighboring communities much worse off due to the resulting water displacement spilling onto their unprotected lands. At some point, the situation might become so bad that the state will need to take over the decision making through the exercise of emergency powers or by amending the state constitution, but when that will happen and at what cost to disadvantaged communities in the meantime is hard to estimate. For many communities, collaborative efforts to protect against sea level rise will likely stagnate until it is too late.

CHAPTER 5

THE PATHS OF WATER
AND ENERGY GOVERNANCE

As discussed so far, extreme weather challenges the infrastructure, practices, and policies that developed and sustained the American West effectively during the nineteenth and twentieth centuries. Droughts have become more frequent and severe, demanding important changes in the way water is allocated and stored, but a legacy of agricultural water rights and fractured water governance perpetuates long-standing inequities and ongoing conflict between farmers, residents, and environmentalists. Wildfires are hotter and more destructive, but efforts at vegetation management and preventing residential growth in risky wildland areas often fall woefully short. Extensive flooding due to rising seas and stronger storms looms over the future of Pacific coast states, but some local communities with ample resources and capacity advantages are moving unilaterally to protect themselves even as they inflict more risk on others.

Some but not all of the blame for slowing progress on climate adaptation is due to the heightened partisanship of contemporary politics related to deeper structural factors such as rising inequality, immigration, global trade, and unresolved racial tensions. Republican resistance to resilience measures derives to some degree from their skepticism about the science of global warming. But some of it, as seen in the surveys, is rooted in the instinctive conservative reluctance to give governments the revenues and authority needed to be effective. However, Democrats also are not as committed to what needs to be done in terms of protecting their communities from natural disasters as they should be. NIMBYism, failure to prioritize long-term threats, and an aversion to new taxes have hobbled resilience measures in predominantly Democratic communities except where disaster relief agencies or insurance companies have finally forced their hand.

Building resilience to extreme weather and decarbonizing the modern economy are two of the hardest challenges modern governments have ever taken on. Politicians are popular when they provide things that people want, not when they tax them for protection they might need in the future. Democracies are ill-suited to getting people to do unpopular things such as giving up their comfortable carbon-addicted lifestyles and refraining from building homes in beautiful but risky places. It is hard to pass and implement farsighted policies when voters expect their elected officials to adhere slavishly to a poll-driven party line that panders to voters' immediate concerns.

While the political and behavioral factors driving climate change policy deservedly receive the lion's share of scholarly and journalistic scrutiny, the role of government structure in shaping policy is more often overlooked. "Compound democracies" divide sovereign power across branches and levels of government in the interests of checking government abuse and ensuring a more deliberative consideration of issues.[1] The unintended effect is that power dispersion feeds policy inertia and enables opportunities for obstruction and delay. The US system is the paradigmatic case in point.

National, state, and local governments in the United States have formally designated responsibilities and powers as laid out in the constitution and elaborated by statues and judicial interpretation. Power is fractured vertically across various levels of government as well as horizontally between different branches of government at each level. The initial purpose for designing the government's core architecture in this way was mainly to prevent centralized tyranny, but over time, governance fracture has increased for other reasons. The federal government has spawned nonpartisan independent agencies (e.g., the Federal Reserve and the Federal Communications Commission) to limit partisan influence over critical areas of policy. Many states have increased the number of separately elected offices to make policy more responsive to public opinion (e.g., elected statewide insurance commissioners) and to better meet the specialized needs of local areas (e.g., special districts for water resource management and fire control).

Geographic boundaries define the limits of jurisdictional authority for state and local governments. For many policy purposes, these traditional divisions work well enough. But the most vexing policy problems, such as building extreme weather resilience and limiting greenhouse gas emissions, often spill beyond the formal demarcations of government sovereignty. To develop shared solutions to regional environmental and economic problems, government officials have to devise cooperative ways to resolve cross-boundary environmental and economic problems. As vertical fracture across federal, state, and local lines has increased, the lines of responsibility have also blurred horizontally as multiple agencies within each level of government work on the same issues. Sorting this out has become even more urgent due to the enhanced dangers of cross-boundary natural disasters such as floods, fires, and droughts.

The blurring of responsibility between national, state, and local governments has been going on for some time.[2] The federal government in 2002 attempted to shape local K–12 education policy, normally a state and local matter, through its No Child Left Behind policy. Similarly, the Obama administration expanded state health care in 2010 through grants and subsidies. States and cities in turn have engaged with foreign governments to forge agreements over shared rivers and aquifers on the Mexican and Canadian borders, tasks that normally belonged to the federal government. There are also many examples of horizontal blurring. Cities have used memorandums of understanding and formal joint powers authority to collaborate with one another on policing as well as fire and flood control, compromising their local sovereign powers in order to solve collective problems more efficiently and effectively.

While these sorts of intergovernmental collaborations have often yielded value, they can sometimes create considerable confusion about who is in charge and ignite disputes over blame when things do not go well. When policy is forged in the same territory by multiple government agencies, there might be no focal point of accountability. Even worse, when the various levels of government are controlled by different political parties, they sometimes work at cross-purposes.

California, the most populated western state, has a massive amount of jurisdictional fracture. The state's executive branch includes over 230 agencies. In addition to the 58 counties and 481 incorporated cities and towns, the state has nearly 3,400 special districts. Special districts are limited-purpose local governments that provide focused public services such as fire protection, sewers, water supply, electricity, parks, and sanitation. About 85% of these special districts are single-function districts that offer one service, and the rest are multifunction districts that offer two or more services. About two-thirds of these special districts have separately elected or appointed boards with directors serving for fixed terms. Another one-third are dependent districts governed by either a city council or a county board of supervisors.

These counties, municipalities, government agencies, and special districts form a complex web of authority throughout the state. When problems such as traffic, sea level rise, and wildfires spill across borders, communities must work out ways to cooperate with one another. In some instances the state uses carrots (e.g., matching grants) to incentivize local collaborations, but in other instances it uses sticks such as regulatory mandates (e.g., clean energy targets) or the threat of taking over if local agencies fail to comply with the requirements of a policy (e.g., California's Sustainable Groundwater Management Act that uses the threat of state takeover to force local areas to develop sustainable groundwater regimes). Intergovernmental cooperation can also emerge from the localities themselves in the form of informal agreements, memorandums of understanding, and joint power authorities. Coordination does not always start at the top.

Improving resilience and decarbonization efforts necessitates better coordination across the region and within a given state such as California. The governance of water and energy systems illustrates this point well. No modern society can operate without adequate water and energy supplies. Water and energy utilities are often described as natural monopolies due to the high costs and necessary scale to deliver water and power.[3] Generating and delivering energy and collecting, cleaning, and then conveying water to households and businesses entail very high costs. Once this infrastructure, such as pipes and transmission lines, is built, it is hard for new entrants to compete except where communities have access to local energy and water supplies. Because natural monopolies can exploit these advantages to maximize profits or neglect their services due to their customers being captured, they need to be held accountable. There are two very distinct forms: investor-owned utilities (IOUs) that are regulated by public utility commissions and publicly owned utilities (POUs) that are accountable to local leaders and electorates.

What makes this water and energy system comparison particularly apt is that while they have followed different pathways to development, in recent years they have converged in form to some degree. Energy systems throughout the twentieth century evolved initially into a more concentrated form, controlled by large vertically integrated publicly POUs and IOUs. Water systems, by comparison, evolved into a highly fractured system with a multiplicity of different local providers and users with various types of legal claims. Over time, the energy system has become more locally distributed due to innovations in local energy generation (e.g., rooftop solar), while the water governance system has inched toward more coordination under state supervision, court decrees, and emergency provisions in state constitutions. Working out the right balance between coordination and self-governance will be critical in order for California and other western states to deal effectively with the dual climate challenges of greenhouse gas mitigation and weather resilience.

The basic poles of intergovernmental relations are between more and less concentrated power at the top of the government pyramid. As a general rule of thumb, coordination is more easily achieved when it comes from the top and is exerted on the units below. Bottom-up coordination, on the other hand, has the advantage of being more inclusive and sensitive to local circumstances. Blending top-down and bottom-up adds to the challenge of vertical complexity. American federalism has also become a more complex horizontal blend of efforts across entities at the same level of government and includes a larger and more diverse array of private and nonprofit stakeholders. There are important lessons in regard to why water and energy systems have converged in form in recent years and what that portends for meeting the political challenges that extreme weather might place on them in the future.

COMPARISONS BETWEEN WATER AND ENERGY

1. Water and energy systems both evolved to meet the demands of delivery, storage, and distribution of a vital resource.

The challenge that settlers faced in living and working in California was how to deliver water from where it was plentiful to the many areas of the state where it was not. As discussed earlier, this involved building an elaborate system of pumps, canals, aqueducts, reservoirs, and dams to collect, store, and redistribute the water. Snow would deposit on the mountains in the winter and then melt in the spring and be used in the dry summer months until the precipitation came back the next winter. When water was in short supply during periods of low precipitation, the diminished surface water supply was rationed to farmers according to a system of appropriative rights, with the first allocations going to senior rights holders (i.e., those who had established the earliest claims in time). Junior rights holders could often offset their surface water losses by pumping water up from the aquifer below their land. Currently, 50% of water use in California goes to environmental flows, 40% to agriculture, and 10% to urban-suburban areas.

In addition, the federal government has provided substantial subsidies to California's agricultural sector since the enactment of the Reclamation Act of 1902. It is estimated, for instance, that 6,800 Central Valley farms receive annual subsidies worth $416 million. Most of this highly subsidized water goes to large commercial farms at a price that is only 2–3% of what residents in Los Angeles and San Francisco pay for their water. In drought periods, this price disparity typically becomes much more salient to the public, causing critics on the Left to question whether the agricultural allocation and subsidies should be reduced and critics on the Right to object to the amount of water devoted to environmental flows.

Like water, electricity is distributed across the state with an extensive system of transmission and distribution lines. Behind-the-meter rooftop solar aside, most renewable and fossil fuel power is generated at the utility-scale inside in the less populated eastern portion of the state or is imported from the regional grid. Electricity is transported by transmission lines to retail service providers and then distributed to customers. As with surface water, California's electricity web crosses many jurisdictional boundaries inside the western region and across the international border with Baja California. These transmission lines are largely owned and maintained by IOUs, such as Pacific Gas & Electric, Southern California Edison, and San Diego Gas & Electric, and two POUs, the Los Angeles Department of Water and Power and the Sacramento Municipal Utility District. Community choice aggregators (CCAs) are a hybrid public form of load serving entity (LSE) that contract for power but depend on the underlying IOU to own and manage any transmission lines.[4]

A critical difference between water and energy is that the grid must always balance supply and demand at every moment in the day, since any imbalance can result in blackouts and service disruptions. The responsibility for this rests with the approximately forty balancing authorities in the western region, the largest of which is the California Independent System Operator. The water system is more forgiving, allowing for emergency water curtailments for agriculture during droughts. Urban and suburban voters are asked to save water but not of course to do without for any long periods of time. Energy shutoffs happen periodically in emergencies such as wildfire events but are rightly treated as politically dangerous by knowing elected officials who remember the fate of Governor Joseph "Gray" Davis during California's energy crisis in 2000–2001.

Another key difference to keep in mind is that major interstate water supplies (e.g., Colorado River water) are allocated by compacts to the upstream and downstream states. The Western Interstate Energy Board market, by comparison, is competitively priced, and the grid load is adjusted to demand. LSEs may contract for a reserve capacity to cover unanticipated surges in demand (i.e., resource adequacy). Dividing water shares in the West is highly contentious, because surface and groundwater supply cannot be manufactured like electricity. Spreading energy transmission across the Western Interconnection area, by comparison, increases grid reliability as a whole by providing access to additional out-of-state supplies of power. The western state grid interconnection will only become more important as states seek to diversify their energy portfolios over time in order to harmonize more time- and weather-variable sources of clean energy such as solar, wind, and hydropower.

2. Water and energy systems spill across international as well as domestic borders.

Water issues have long been a source of contention between California and Mexico. One issue that dates back to the nineteenth century is apportioning Colorado River water shares between the two countries. The Treaty of 1944 gave Mexico a guarantee of 1.5 million acre-feet of water and potentially more under certain circumstances. The treaty also established the International Boundary and Water Commission as the agency tasked with addressing and overseeing the resolution of binational water disputes. The commission has the ability to adopt minutes that have the force of treaty and can address issues as they arise such as water salinity, the construction of new dams along the border rivers, and the design of drought measures. There have been 170 minutes adopted over time.

Most of the controversy and negotiations to date have been between Texas and Mexico, but the California–California Baja issues are equally contentious. Positioned at the end of the line after water is diverted down the All-American Canal to California, Mexico receives a much smaller share of poor-quality water. When California built the All-American Canal to transport Colorado River water to San Diego, it was lined in such a manner as to adversely affect the aquifer that California Baja shares with San Diego. With water scarcity likely to become even more problematic given climate change in Mexico and the US Southwest, Colorado River water disputes will also become more contentious. California Baja already suffers from severe water shortages, and climate projections suggest that it could receive 60% less precipitation in the future. California Baja and San Diego have both constructed large desalination plants near the San Diego border. Given the problems with the outflow of residual brine and the incidental intake of biota, both sides of the border will need to coordinate the operation of these and future plants to minimize these problems.

Another issue is industrial pollution from the Mexican side of Tijuana River. The river flows north into San Diego, and the pollution it carries is treated by a waste treatment plant in San Diego before it is released into the ocean. California would like Mexico to do more to reduce the river pollution before it enters US territory, but there has not been much progress on the issue.

As defined by the La Paz Agreement between the United States and Mexico, the US-Mexico border region is a zone stretching one hundred kilometers on either side of the international boundary. This agreement covers the California–Baja California border region. The one hundred–kilometer (sixty-two–mile) zone includes all of San Diego and Imperial counties and the Mexican communities of Tijuana, Rosarito, Ensenada, Tecate, and Mexicali. This area encompasses all main population centers of the region and contains its principal energy-related infrastructure.

The primary energy connections along the border are imported natural gas and renewable energy. Baja California does not produce any of its own natural gas and relies entirely on imports, mainly from the United States. During 2019 Baja California consumed about 340 million standard cubic feet of natural gas per day, with most of the energy going toward industrial uses. Although the region has few nonrenewable deposits, it does have a lot of potential for wind and solar energy, given strong wind patterns and high solar irradiance there.

A wind farm was built in Baja California based on a power purchase agreement by San Diego Gas & Electric for $820 million over twenty years in 2015. This 1,200-megawatt plant was comparable to that of the largest wind plant in the United States, giving Mexico a bigger footprint in the energy market and providing the surrounding communities with essentially passive income from the profits of renting out the land to a private energy company. Siting wind farms in Mexico is easier than in the United States due to laxer environmental standards with respect to protecting migrating birds from the turbines.

The Western Interconnection is the energy grid that spans more than 1.8 million square miles in all or part of fourteen states, the Canadian provinces of British Columbia and Alberta, and the northern portion of Baja California in Mexico. The Western Interconnection is governed by the Western Electricity Coordinating Council. Baja California, a member of the council, has two international power connections clusters with California, one at the Tijuana-Miguel border and another at the La Rosita–Imperial Valley border. Baja California's international gas border crossings, a crucial component of its energy portfolio, comprise a combined volume of two billion cubic feet of natural gas per day. Firms that have invested in the pipelines include San Diego Gas & Electric, Ecogas, Sempra, TransCanada, and the Southern California Gas Co.

In 2014, Governor Jerry Brown signed a deal to enhance cooperation on energy and the environment with Mexico. The agreement pledged to share information about science, technology, and regulatory policy related to decarbonization; address air-quality problems on the border; and to facilitate the transition to zero-emissions vehicles. Baja California has experienced strong population growth and economic development, which has generated more water and air pollution. While there was some follow-up and exchanges on these items, the elections of President

Donald Trump and President Andrés Manuel López Obrador essentially halted further progress on this initiative during their terms in office. Baja California imports a substantial amount of gas from the United States and is tied to the Western Interconnection, but aspirations for a tighter connection to California's climate change have been slowed considerably by political changes at the national level in both countries. Still, Mexico has committed to ambitious climate goals and efforts to promote greater collaboration between California and Baja California that will no doubt resume at some point when the political conditions are right.

3. The water system historically has been more fractured than the energy system.

Water in California is regulated primarily by three state agencies: the Department of Water Resources, the State Water Quality Resources Control Board (with nine regional counterparts), and the California Public Utilities Commission. There are also sixteen other state agencies with regulatory authority over various aspects of water governance, including the State Water Commission, three agencies that deal with the Delta, and two that deal with the seacoast.

The number of state-level governmental water agencies pales in comparison to the number of local water entities. There are approximately 2,900 separate community water systems falling into twenty-six distinct types. They include publicly owned systems (e.g., cities and counties), independent special districts, state and federally owned systems, investor-owned utilities, other private systems, and more. These local water entities often have very different institutional designs. A study done in 2002 identified over 1,200 water districts with statutory authorizations.[5] These water districts have different types of governance structures. About a third of these water districts employed a dependent governing body that is directly controlled by either a city or a county. Dependent districts are governed by a city council or county board of supervisors or by city- or county-appointed representatives. The other two-thirds of water districts employ independent special districts where the governing body is either directly elected by voters or appointed for a fixed term of service by a board of supervisors. These institutional design variations create different bureaucratic dynamics, constituency pressures, and accountability issues.

This agency heterogeneity is less problematic during a drought emergency because the governor has fairly sweeping emergency powers. It is more problematic when there is no immediate emergency but nonetheless a critical need to prepare for the next drought. The state's Integrated Regional Water Management program made a valiant effort to bring local water entities together to develop joint projects and planning at the watershed scale but with mixed results.

The landscape of entities that control electricity in California is not as complex and varied as in the case of water. But it is certainly not simple. Energy policy has been made by executive actions, normal legislative processes, agency discretion, and bargaining with stakeholder groups. The same could be said for aspects of water policy, but the state has much firmer regulatory control over the electricity market than it does over water supply and usage.

Several state agencies play a significant role in energy policy. Each has a defined role. The California Energy Commission oversees planning and has particular responsibility for the POUs. The California Public Utilities Commission oversees the other retail sellers such as the IOUs and CCAs. The California Air Resources Board regulates mobile sources of pollution and greenhouse gas emissions and is most heavily involved in the effort to increase electric vehicle use and build more charging infrastructure. The California Independent System Operator is an autonomous entity that operates a competitive wholesale electricity market and manages reliability for 80% of the transmission grid that the LSEs rely on.

The state legislature and recent governors have set out very ambitious decarbonization goals, and these agencies have done an admirable job to date implementing them. They also have made a strong effort to collaborate with one another, particularly in the context of setting out the state's Integrated Resource Plan. But the open multiparty process of negotiation between these agencies and stakeholder groups is laborious and time-consuming.

The rise of CCAs has introduced new tensions over who pays for legacy power contracts and how much excess energy must be purchased to maintain system reliability. Essentially, these new entities allow cities by themselves or in collaboration to control their local energy portfolios, enabling them to purchase either cheaper or green power supplies. Adding these cities to the mix has facilitated a more rapid statewide transition to green energy to some degree but at the expense of complicating the state's task of overseeing an orderly transition away from fossil fuels. The CCAs are at loggerheads with the state over several matters such as their responsibility for resource adequacy, the length of the power purchase agreements, and paying for residual contracts and wildfire liability damages. While this has complicated energy planning in important ways, it is still fair to say that the state has a more coordinated energy system than water system.

By comparison, the California Water Plan is mainly an impressive compilation of data and information about state water resources.[6] The Department of Water Resources manages the state's water network infrastructure effectively and has a hand in overseeing the Integrated Regional Water Management and Groundwater Sustainability Agencies programs. The State Water Resources Control Board, along with the governor's office, played a central role in designing drought policy in the critical 2012–17 period. But the state has far less control over water usage in nonemergency periods.

As a consequence, the state continues to build new housing and commercial developments in water-stressed parts of the state. Some new housing developments in southern California include artificial lakes and golf courses, apparently on the assumption that they will have continued supply from the Colorado River despite climate change and population growth in upriver states such as Nevada, Arizona, and Colorado. At the same time, Central Valley counties permit farmers to plant fruit and nut trees that cannot go fallow in dry years and depend on increasingly depleted groundwater supplies.

These water governance problems foreshadow some of the local obstacles to state designs that California will encounter when it tries to decarbonize the economy by reaching into matters that are normally controlled by local governments, such as requiring more housing density along rail lines to cut down on car emissions and eliminating end-use gas in family and commercial residences. Local governments carefully guard their sovereignty over local roads, economic development, and zoning and tend to resist state interference. They are particularly resistant to unfunded regulatory mandates that require them to make changes at their own expense. Most of the successful state efforts to enhance local water resource capacity building have relied instead on public bond funding and matching grant programs that incentivize rather than require local cooperation.

A core water task is to substitute old forms of surface water storage with stormwater reuse, desalination, and aquifer replenishment. In addition, the power for the pumps that extract, clean, and store water has to become greener. That electricity sector challenge is quite formidable: the state needs to shift its energy profile from fossil fuels to clean energy and meet benchmarks set by the governor and the legislature. These goals have become progressively more ambitious targets over time with respect to lowering emissions as compared to 1990 levels and achieving carbon neutrality by 2045. The initial benchmarks were set by legislation, but Governor Brown used his executive authority to up the ante for decarbonization. Much of the progress to date in meeting these goals is due to regulatory pressure on the LSEs to meet renewable portfolio standards. In addition, the state has tried to encourage voluntary green incentives (i.e., renewable goals above the renewable portfolio standards' floor levels) through its CCA program.

While the record of achievement to date has been impressive in many regards, there are some concerns that arise from tensions between the various new state energy goals and the necessity of maintaining a balanced portfolio of energy options in order to deal with steep evening ramp-ups and prolonged heat waves in the western region. Both can stress the available power supply.

The state's current renewable energy definition precludes certain forms of clean energy such as large hydro and nuclear, which many LSEs are currently counting toward their clean energy target. It is not clear whether the state can build enough utility-scale solar and wind stations to offset the loss of these nonrenewable categories of clean energy. In addition, the California Independent System Operator has concerns about the rate at which the state is retiring gas plants and the impact of such a fast build-out of solar without adequate levels of storage to ensure grid reliability. The ongoing fight over CCA exit fees also manifests a disagreement between local community LSEs and the state over who assumes the burden of legacy energy contracts and wildfire liability costs.

Electricity regulation was much simpler when the state could work through the large POUs and IOUs, but the proliferation of CCAs in recent years has raised the transaction costs (i.e., the time and effort devoted to negotiation between agencies and stakeholder groups) and lessened the state's control over retail purchases. It is important to bear in the mind that the electricity system represents only 15% of the state's emissions. To make a serious dent in that overall emissions goal will require taking on transportation and industrial sectors of the economy, efforts that will put even more pressure on state-local government relations. But given that the long-term plan is to electrify many more things in the future, the success of the decarbonization effort hinges crucially on greening the grid as quickly as possible. The rationale is pretty straightforward: if the state moves forward to electrify the transportation sector but depends to some significant degree on amounts of imported power generated by gas or coal in order to charge vehicles at night, this will undermine the state's overall emissions goal even as it expands the number of electric vehicles.

4. Both the water and energy systems are under pressure from climate change.

Aside from the impact that severe droughts have had on surface and belowground water supplies, large wildfires can burn watershed areas, potentially affecting both the supply and the quality of water. Human activity is the greatest source of wildfire ignition. As people continue to build and live next to or within wildland areas, the risk of fire ignition increases at a time when forests are drier and more susceptible to horrific fires. An important water-energy connection is evident here: construction in new areas requires new electricity transmission and distribution lines, which raises the odds of more ignitions and potential harm to vital watersheds.

Wildfires can cause extensive damage to watershed water quantity and quality, snowpack, debris flow, and aquatic ecosystems. The damage to vegetation increases the overland flow, leading to soil erosion and debris flow and adding potentially harmful sediment and contaminants to the water. The sediment flow can make its way into reservoirs, which decreases their storage capacity and shortens their life spans.

The resilience problems also pose significant challenges for water resource management. Planning and building water infrastructure requires anticipating and adjusting to extreme weather threats but with a fair degree of uncertainty about where and when such weather events will occur. For many decades, water planning could assume that there would be year-to-year variations in extreme weather but a relatively stable climate. Now, water infrastructure planning must also account for changing climate conditions that will likely increase the frequency and intensity of extreme weather events such as drought and fire. The data about water conditions in the state exist in abundance, but the effort to improve planning about water resources on a watershed scale had a shaky start in the state's Integrated Regional Water Management program. The local entities managed to organize themselves more or less around watersheds and collaborated to some degree to apply for matching grants that the states offered, but in many cases when the money dried up, so did the collaborative efforts.[7] The capacity of the state or indeed regional entities to get local communities to operate collaboratively when the state is not in an emergency situation is still too weak for the likely extreme weather challenges ahead.

The electricity system is a cause of wildfires but is adversely impacted by them as well. When the Diablo and Santa Ana winds are blowing hard, transmission lines can fall or be struck by trees and spark a fire. As a precautionary measure, the grid operators can call for public safety power shutoffs that stop the flow of electricity to businesses and residences until conditions improve. If the danger persists for several days, safety power shutoffs can be very disruptive for large numbers of customers. Shutting off power for safety reasons may be even more disruptive in the future if the state electrifies everything to move away from dependence on gas and fossil fuels. Complicating matters further, wildfires in other states can affect California's energy supply due to the western grid interconnection. Wildfires are a regional problem.

Organizations that study and make recommendations about resilience problems at the regional scale include, among others, the Western Governors' Association, the Western Electricity Coordinating Council, and the Western Interstate Energy Board. There are ways to improve vegetation management around transmission lines. Sensors and dynamic line rating systems can identify transmission problems more quickly. And there are more efficient ways to cut off power during wildfire emergencies while minimizing customer inconvenience and economic loss. But taking these steps is largely left up to the utility companies and is not controlled by any single government entity. Once again, we run into a core intergovernmental question: To what extent should coordinating organizations have stronger means to either influence or compel state and local entities to undertake measures that could enhance electricity reliability and resilience overall?

5. Both systems are moving toward developing new resources and more storage.

The seasonality of California's weather means that the state has needed to find ways to store water in the winter, when it is most likely to rain, for use in the summer when precipitation is rare. As discussed earlier, this means building not only dams and reservoirs throughout the state to store the winter water but also aqueducts and canals to convey it. That infrastructure is now aging, prone to breaking down (e.g., the recent damage to the Oroville Dam spillway), and is harder to replace due to concerns about environmental impacts and costs. The drought in 2012–17 took its toll and reduced water levels in the reservoirs. The water in California's reservoirs has in recent years been well below both capacity levels and historical averages. While the variability of western weather means that there will be rainy as well as dry periods, states throughout the region will need more storage to endure longer droughts until the rain returns. Groundwater, the backup when surface water is scarce, has dropped significantly over time in many places due to overuse and low precipitation. While California has made more effort in recent years to replenish its depleted aquifers, water subtraction still exceeds natural and mechanical replenishment, creating a cumulative water deficit.

At the same time, it has become harder politically to build reservoirs and dams. The last large reservoir constructed in California was the New Melones on the Stanislaus River in 1979. Dam construction has also slowed to minuscule levels, while some dams have been or will be torn down either because they have filled with silt or in order to restore the ecology of the area that was flooded when the dam was constructed. As a consequence, California water policy is turning toward recycled water and desalination, despite the higher cost and personal aversion some people have to drinking treated water.

For most of the twentieth century, the power companies' goals were to provide reliable energy at the cheapest cost. Now, there are additional requirements such as switching to clean energy, ensuring the resilience of the transmission system to wildfires, and building storage and resource adequacy to offset the novel demands of extreme weather. Greening the grid is an essential first step, because many of the next decarbonization steps will increase the total demand for electricity.

Emissions from the electricity sector only contribute 15% of total carbon emissions in California. Transportation alone accounts for 41%, which is why the state has committed to getting five million zero-emissions vehicles on the road and building ten thousand DC fast-charging stations. But building out charging stations requires siting the chargers strategically, acquiring the land, and making it through the state's cumbersome environmental permitting processes. The California Air Resources Board can provide funds and incentives toward these goals, but achieving them will require working with local government officials and dealing with inevitable objections from some residents. The CCAs could possibly facilitate this process, as many of them have programs to promote zero-emissions vehicle use, but CCAs vary enormously in their commitment to decarbonization and their capacity to do anything about it.

Beyond the transportation sector, decarbonizing the agricultural and manufacturing sectors will be costly and could place California firms at a competitive disadvantage with companies outside the state. Engineers can propose lots of possible technical ways to lower the carbon footprint for each of the sectors, but many of these ideas carry large costs. Methane emissions from livestock could be reduced with anaerobic digesters. Carbon capture and storage could reduce emissions from cement production. Electrifying trucks and fueling them with green hydrogen could lessen the carbon footprint for many businesses. But all of these measures would raise the costs for the companies and make them less competitive with firms in red states. There are market constraints to what California can achieve on its own if other states do not also decarbonize at the same time.

6. Water and energy policy are matters of equity, not just efficient allocation.

Poorer communities tend to have bad water resources. This is particularly true for rural communities in the Central Valley and along the Central Coast from Santa Cruz to Santa Barbara that are dependent on groundwater supplies. As discussed earlier, groundwater has been depleted over many years in these areas as a consequence of withdrawal during drought years and inadequate aquifer recharge. Often, poorer communities have to deal with both quality and quantity issues. Along the coast, the water-quality problems associated with chemicals and nitrates are compounded with saltwater intrusion and sea level rise. The Coastal Commission, despite its strong powers, is largely focused on limiting development and preserving beach access. Mostly, water-quality problems are dealt with by the State Water Resources Control Board.

Even outside the problematic areas, there are water justice problems. Some disadvantaged communities rely on small mutual water companies that lack the capacity to install money-saving efficiencies and infra-structure improvements. During droughts, higher-income people can cut their usage by letting their lawns go brown, whereas poorer people have to cut back on indoor water uses for bathing, cleaning, and cooking. The communities that could benefit the most from programs that would subsidize water costs for the poorest residents lack the revenue to institute such subsidies, the capacity to enact the surcharges that could pay for them, or both. And water-quality problems abound in disadvantaged communities. In the Central Valley, farmworkers have to rely on poor-quality groundwater supplies or bottled water if it is provided or they can afford it.

The increasing diversity of the legislature has helped to some degree. The minority caucuses have grown after the last three rounds of redistricting and have tweaked the criteria for grant awards to reward proposals that include disadvantaged community areas. But poorer cities often lack the internal capacity to go after technical water grants and the private funding matches that are often required in the funding competitions.

In the case of energy, a core problem is the very high energy costs in the state. As a consequence, state policy with respect to energy justice relies heavily on subsidies for low-income energy consumers offered through two programs that are administered by the IOUs and CCAs: the California Alternate Rates for Energy Program, which offers a 30–35% discount on electricity bills and 20% on natural gas, and the Family Electric Rate Assistance Program, which offers a monthly discount of 18% on electricity. POUs have their own energy assistance programs. Residents who are struggling to pay their utility bill can qualify for the Low Income Home Energy Assistance Program assistance, which pays up to $1,000. These programs are important because California's electricity rates are higher than in the country as a whole.

Another set of equity problems regarding electricity concerns the inability of disadvantaged communities to participate in rooftop solar and zero-emissions vehicle programs. This means that they cannot participate as fully in the green economy, and they also miss out on the savings from participating in net metering programs. Analysis at the county level shows that high-income communities are more likely to receive solar subsidies.[8] Much the same applies to zero-emissions vehicles and hybrids. One study found that 70% of electric car owners earned at least $100,000 per year.[9] Other studies show that owners of hybrids and electric vehicles tend to be better educated, between the ages of thirty and forty-nine, and disproportionately male.[10] Designing these programs so they reach more of the disadvantaged communities is a matter of not just fairness but also scaling up: we cannot widen decarbonization efforts without moving the effort down the socioeconomic ladder.

THE CURRENT STATE OF WATER AND ENERGY SYSTEMS

Securing water supplies under conditions of more extreme weather is a resilience problem. Decarbonizing the grid is a foundational element of the strategy to mitigate greenhouse gas emissions that are causing global warming. Water and energy policy have both become more complex over time. In the nineteenth century, providing water resources in a predominantly arid and semiarid region was instrumental to population and agricultural growth. With the emergence of the environmental movement in the mid to late twentieth century, opposition to large dams on sensitive natural habitat grew, and federal financial support for construction dwindled. New water solutions have to be more environmentally friendly as well as more equitable and inclusive.

California's electricity system was initially set up to provide reliable, affordable power for a growing population and commercial sector. In recent decades, new goals have emerged such as transitioning to clean power, providing resource adequacy to compensate for the intermittency of weather-dependent renewables, and ensuring fairness for low-income groups. As in the case of water, the competing goals have multiplied over time, forcing governments to find balanced approaches.

In both cases, the legacy practices that at one time functioned usefully are more problematic today. When the West was being settled, agriculture and mining were the main commercial enterprises. To foster agriculture, the government conferred water rights and subsidies on farmers. The federal government's 1902 Reclamation Act, for instance, promoted irrigation in the arid areas of the West by constructing dams and reservoirs. By reducing repayment provisions and lowering prices for farmers, water became heavily subsidized and still is today. States did their part by conferring water rights to farmers and promoting irrigation. California's appropriative water rights system prioritized water allocations to farmers based on the date of the claim and continued use of that allocation. And the state-built water infrastructure that transported water from the north to the semiarid areas of the south with the successful passage of the State Water Resources Development Bond Act of 1960 (i.e., the State Water Project) enabled extensive irrigation in the Central Valley.

Because individuals, companies, and communities could lay claim to the water right under on their land, California ceded control and coordination of groundwater use, which led to overdrafting problems in the late twentieth century. Following market forces and absent any central land-use control, farmers gravitated away from growing cheap crops that could be fallowed in drought years and moved toward planting higher-value fruit and nut trees that could not be fallowed. When surface water was cut off during drought years, farmers then relied on groundwater resources to back up the missing surface water allocations. Over time, this led to critically overdrafted groundwater basins. While California has a state water plan process and has enacted legislation that incentivizes more collaboration (e.g., the Integrated Regional Water Management Act and the Sustainable Groundwater Management Act), there is still no central coordinating agency that rationalizes water use in the agricultural and housing sectors. Farmers continue to plant more orchards, and developers continue to build more housing in water-stressed areas of the state.

The electricity story is one of California moving toward more institutional decentralization and losing some of the policy coordination advantages it enjoyed throughout most of the post–World War II period. The system evolved around large vertically integrated IOUs and POUs regulated by the state legislature, the California Energy Commission, and the California Public Utilities Commission. The demands of balancing the grid and of financing the costly long-term investments for power generation kept the system from becoming as fractionalized as water delivery was. The regulatory goals were relatively straightforward as compared to today: ensure that these utilities would not exploit their market power for excess profit, and keep the lights on reliably.

The urgency of greening the grid and the possibility of more distributed systems of energy generation with associated battery storage have pushed the energy system in a more decentralized direction. In Marxist language, innovations in the means of production are shifting the relations of production. Additionally, multiple goals have to be balanced: decarbonization, energy equity, resilience to extreme weather, resource reliability, and minimal ecological harm.

This would be hard enough if the major state agencies had retained total control over the electricity purchases, transmission, and distribution, but energy reforms in the twenty-first century have broken up vertical integration and opened up new hybrid forms of LSEs such as the CCAs.[11] Most CCAs are local governments, singly or in a joint powers arrangement, that purchase power for their communities while relying on the transmission and distribution systems in their IOU territory. CCAs vary greatly in their technical capacity and their dedication to decarbonization goals: in wealthier liberal communities the CCAs are rushing ahead to establish 100% clean energy portfolios, while in other less-advantaged communities the CCAs are struggling to stay afloat financially and still meet the state's minimum renewable portfolio standards.

Letting affluent first-mover communities purchase renewable power contracts in excess of the state's baseline renewable portfolio standard adds more renewables to the overall energy mix. On the minus side, this has lessened the state's coordination over the transition to clean energy, potentially undermining the system's reliability and resilience. As in the case of sea level rise, affluent local communities left to their own devices can impose negative consequences on the disadvantaged. In the quest for ostensible energy purity, communities that launch their own LSEs try to evade the residual costs of older energy contracts and wildfire liabilities, leaving the lower-income communities who remain with the IOU to pay. This leads to ongoing tensions between the state and the local government LSEs over so-called exit fees (e.g., payments added to utility bills to pay for older long-term expensive energy contracts, resource adequacy requirements, and the fixed costs related to wildfires). The simpler, even if imperfect, vertically integrated system has given way to a more decentralized system, with local agencies wanting to go their own way and resisting central control. Electricity governance is gravitating toward the water model even as the water system attempts to move in a coordinated direction. In short, water and energy are converging to some degree from their very different starting points, attempting to balance the benefits and costs of more centralized coordination while retaining the benefits of more distributed systems.

THE INTERSECTION OF SYSTEMIC CHANGE
AND MORE EXTREME WEATHER IN THE FUTURE

The forms of government shape policy in many ways. When state and local jurisdictions have primary policy responsibility in a given area, the federal government has to use indirect instruments to induce compliance unless some regulatory hook allows it to compel states and localities to undertake certain actions by virtue of owning land in the state or as a condition of receiving federal support. But policy conditions can in return shape the forms of government. Emergencies can give more leverage to particular branches of government (i.e., governors and presidents over legislatures) and to one level of government over others (i.e., to the state over local government due to emergency powers during natural disasters and public health crises such the COVID-19 pandemic). Emergency powers are typically given conditionally in the United States.[12] The premise is that they should only last as long as the crisis. But what happens if the crisis is ongoing? Contingent emergency power could become embedded as a structural feature. If extreme weather becomes more frequent and extreme, emergency powers could become normalized, allowing governors and presidents to override normal processes of consultation and legislative approval.

An alternative to this is to work out stronger mechanisms of collaboration among the various actors. In the energy realm, this might mean giving more power to organizations such as the Western Interstate Energy Board and the Western Electricity Coordinating Council to coordinate the grid across the region to allow for more ways to compensate for the intermittency of renewables or to harmonize preventive actions and emergency processes that would minimize power shutdowns during extreme weather situations. With respect to water, the Colorado River illustrates the potential value of anticipating future water shortages during prolonged and more severe drought conditions and developing a more coordinated strategy for sharing and conserving water along the river. The thick web of interstate treaties and legacy water rights presents a major obstacle to regional water coordination. It is hard to see much hope on that front at the moment. It is more likely that the water system will try to postpone the day of reckoning with greater emphasis on using water more efficiently and trying to recycle water, but it is possible that down the road, conditions might deteriorate to a point where Congress will have to take on the issue of interstate water shares much more seriously.

CONCLUSION

A More Resilient West?

The American West is a region that has always posed difficult challenges to would-be inhabitants due to its climate and topography. Global warming has added rising seas and early snowmelts to the mix but for the most part has made and will continue to make familiar western problems worse, among them, harsher droughts, hotter wildfires, stronger Pacific storms, and more extensive flooding in low-lying coastal areas and along inland rivers. Familiarity with these kinds of weather events and memories of what worked in the past can breed comforting but ultimately costly fictions, such as believing that weather patterns will return to a normal range because they have always done so before or that there is no need to change human behavior because technology will save us eventually. But ignoring the need to build resilience now will only add later to the cumulative damage inflicted by extreme weather, leaving future generations to pick up the tab.

Science enabled global warming. Industrialization was founded on fossil fuel technology, making life easier and more comfortable but all the while emitting greenhouse gases into the atmosphere. Because turning back the clock on modern living is unimaginable for numerous social and political reasons, we have chosen to green the power mix instead. Investing in resilience measures will both protect us from the consequences of past emissions and hopefully buy enough time to decarbonize the world's economy.

A central theme in this book is that the political challenges with respect to climate policy are at least as challenging as the technical ones. Indeed, it is quite possible that the politics will lag behind the rate of technical innovation until there is no other choice. Technically, we have to develop enough clean energy and storage to replace fossil fuels and at the same time find innovative, affordable ways to protect people and property from harsher weather. But these will not get enacted unless there is sufficient buy-in from elected officials and the public. As long as people and businesses continue to move into risky wildland and low-lying coastal areas, the expense and difficulty of protecting them will continue to escalate.

The COVID-19 pandemic has taught us a lot about the ways human cognition and institutional features mediate policy solutions. Trust in the messenger now matters as much or more to many Americans these days as the substance of the message. Many Republicans seem to have more confidence in their leaders than in scientists and public health experts. This has led them to question the pandemic's seriousness and to resist wearing protective masks. Their skepticism of medical expertise has fused with long-simmering resentments of coastal elites and universities and has hardened with the experience of the pandemic. Not wearing a mask became an emotionally fraught political and social statement, signaling solidarity with like-minded friends and family. It might also be a harbinger of future climate change politics.

Decarbonization in particular could easily follow the same path of popular resistance. Going green disrupts traditional economic activities related to fossil fuel extraction that have sustained interior portions of the West for many decades. Defending coal interests has taken on a symbolic importance that far exceeds the economic consequences of closing down coal mines and replacing them with wind turbines and utility-scale solar. The residual antiquated features of the American political system have played a role as well. The constitutional design of the US Senate and the presidential electoral college combined with the filibuster give disproportionate voice and power to those who want to resist change and even turn back the clock on environmental policy.

But working across party lines on resilience may someday play an important role in softening the partisan divide on climate policy, just as the hospitalizations, deaths, and long-term COVID-19 symptoms may gradually whittle away red state resistance to pandemic policies. Minus extreme weather, climate change is fertile ground for symbolic warfare. Chemical reactions in the atmosphere without physical manifestations on the ground are abstract possibilities, not experienced realities. Unable to judge for themselves, people rely more on trust in the messenger or listen only to the messages they want to believe. This may change eventually if the climate consequences become bad enough and more directly impactful. The messenger becomes less important when you have your own data. People begin to trust their lived experience more and the opinions of vote-seeking politicians less. But getting on the same page about climate policies will take time. People have to make the connection between the harsher weather events and the abstractions of atmospheric chemistry and also between the higher home insurance rates they are paying and the social consequences of allowing people to inhabit high-risk areas.

POLITICAL LESSONS

The experience of trying to cope with new extreme weather circumstances has taught us a few valuable political lessons to date. The following is a review of some of them.

1. While partisanship polarizes public attitudes toward resilience policies to some degree, there are nonetheless some important points of bipartisan agreement to build upon.

Polarization matters, but we have seen that it is more complicated with resilience than one might realize. Republicans are less likely than Democrats to report that they have experienced extreme weather events such as wildfires and floods even when objective evidence suggests otherwise. And while the partisan gaps on resilience policies are not quite as extreme as they are on issues such as abortion, they are strong enough to stop some people from taking the proper precautions when wildfire smoke reaches unhealthy levels or from voting for raised taxes in order to build higher levees.

Moreover, some of the resistance to resilience projects emanates from the other side of the partisan divide. Water storage has always been and will become even more critical in the American West. Some forms of storage are more controversial than others, but objections to building additional dams, for instance, largely emanate from the liberal environmental, not the conservative end of the ideological spectrum.

Still, there are also hopeful examples in recent years of Democrats and Republicans getting on the same page when they have to, such as voluntarily accepting water cutbacks during drought emergencies and supporting new water-supply projects that recycle water, replenish aquifers, and desalinize seawater. Democrats and Republicans also agree that people and communities should take steps to protect themselves from wildfires and floods, a sentiment that is reinforced by living through a natural disaster. But projects that might potentially protect communities sometimes get lost in the translation between wanting something and actually doing something about it. In some cases, such as droughts and Pacific flooding events, it is because the weather is not yet extreme enough. In other cases, such as wildfires, it is because the extreme weather threat is either too localized (e.g., not enough voters live in the high-hazard wildfire severity zones) or because its hazard is underestimated (e.g., the adverse health consequences of wildfire smoke).

As a consequence, all too often voters reward their political officials for funding relief measures but not for prevention. In a democracy, politicians stay in office by doing what the voters want, not what they should want. Wildfire smoke reaches a much wider swath of the population, and the medical evidence about its many adverse health effects is mounting. But as with secondhand tobacco smoke in the 1970s, the harms are not yet widely enough recognized to have political impact. This could change in the future as it did for secondhand tobacco smoke in the past.

2. The increasing collective costs associated with natural disaster damages and rising insurance rates may be more impactful in the future than additional climate shaming and sermonizing above current levels.

As the COVID-19 pandemic has taught us, it is harder now to build broad political consensus than it was in the immediate post–World War II period. You can get part of the way by educating people about the personal costs of inaction, a little further by applying social pressure, and further still with tax credits and subsidies. But inevitably some people will resist to the bitter end. A key consideration for policy makers is whether the benefit being offered is excludable or not. In the decarbonization case, the benefit of achieving a manageable global warming level is not excludable, and thus free riding (i.e., getting the benefit but not paying for it) is a serious problem. Given the gravity of the climate change problem, it is quite likely that governments will eventually have to compel people with regulations and penalties to conform. Needless to say, this will be a bitter political battle.

In the case of climate adaptation measures, however, it is possible to build community protection from many types of weather harms in an excludable way. Raising housing and buildings above the expected waterline and constructing high levees are good examples. The degree to which a benefit is either severable or inextricably shared shapes political decision making. When it is severable, the political path is easier, allowing politicians to avoid the messy business of compelling voters to do things they do not want to do. If voters want the protection of a levee, for instance, the community can move ahead unilaterally and protect itself. If not, it can let things stand as they are. When the benefit is not severable in this way, there is no easy path forward. Such is the more difficult politics of mitigation, as it requires not just cooperation between entities in any given country but also collaboration across the globe.

That said, resilience choices involve political risks that can stymie policy unless there is a forcing hand. The risk of the do-nothing strategy is that the community could lose its bet with nature, and people in charge will be blamed. Given the odds of a one hundred–year natural disaster versus the odds of a taxpayer revolt in the next election if taxes are raised to pay for costly projects with counterfactual benefits, many elected officials may choose to accept the weather risk over the political risk. The situation changes when the costs of doing nothing become immediate and concrete. A politician representing constituents who need a levee in order to prevent a substantial hike in their insurance costs is one example of a forcing hand. If the cost of the levee is then put on the credit card (e.g., a bond measure paid off over a long period of time), the political calculus greatly favors the levee choice over doing nothing and being responsible for the higher insurance rates. The costs of natural disasters turn hypothetical harms into immediate concerns, which is what the political system is better at reacting to.

By comparison, the possible harms that displaced water impose on neighboring communities due to unilateral action pose little or no political threat unless the local political official in charge envisions running for a higher office that includes the neighboring communities. Presentism will win out more often than not. The easiest political pathways are not always socially optimal in the resilience case. Without a collaborative approach, the vulnerability gap will grow between disadvantaged and well-resourced communities.

3. Climate policy inequities can arise sometimes unintentionally but can only be fixed intentionally.

Given the history of redlining in western states (i.e., exclusionary pacts that kept nonwhite and disadvantaged groups from living in white neighborhoods), there are many instances in which poorer populations live in places that are prone to air pollution problems, toxic soils, impure water, and the like. Some of them are also at risk from floods and wildfires or live in homes that lack the air filter systems and home insulation to protect from toxic wildfire plumes.

As we saw with the Foster City example, communities with resources and capacity advantages can build levees to shield their properties from future sea level rise. But doing so displaces water onto neighboring communities. There are morally principled reasons for not letting things evolve this way, but there are also long-term social interest considerations. When people living in low-income housing are displaced, they are rarely insured or able to rebuild. Leaving aside the economic damages they suffer, they are then forced to move out into faraway less expensive exurbs and then commute back to work into the already traffic-congested urban areas. These longer commutes often extend beyond the range of mass transit and the current range of electric vehicles that consumers will soon be required to buy. And until we achieve the scenario zero-emission vehicles, longer commutes in gasoline-powered cars and trucks means higher transportation emissions.

Whether for reasons of compassion or self-interest, success in both decarbonization and resilience requires addressing issues of equity and inclusion. Scaling up the clean electrification of the grid and other sectors of the economy eventually means extending these efforts down the socioeconomic ladder. Affluent communities that jump ahead can serve a valuable purpose by pioneering new technology. Hitting our emission targets and limiting the social costs of more extreme weather require much greater population outreach. Ignoring such considerations is analogous to not vaccinating essential workers and third world countries during a dangerous pandemic.

4. Beyond partisanship, there are other problems—such as NIMBYism, a failure to recognize trade-offs, virtue signaling, and unwillingness to pay—that must get resolved to develop better extreme weather resilience.

The conventional story about climate change politics identifies polarization as the core problem. But a closer look suggests that quite a few problems aside from the partisan gap retard climate progress even in solidly blue areas. On the emissions front, NIMBYism blocks the widescale deployment of utility-scale solar and wind projects in open spaces near upscale blue neighborhoods and pushes them to remote areas. This is not just a trend in the American West. Massachusetts communities buy most of their wind power from Texas facilities, just as upscale California communities in Marin and Sonoma in California contract for green power generated by utility-scale solar facilities in the eastern rangelands.

In this way, affluent liberal communities get to proclaim their renewable virtue while the conservative areas get profits from the sales. Such complementary trades are desirable in many ways except for one: they require stringing new transmission and distribution lines in order to carry the power from where it is sourced to where it is used. There are several problems with this. Siting transmission lines is often politically problematic because people do not like having them near their homes unless they are buried, which is expensive. And as discussed in chapter 2, downed power lines can ignite wildfires that release carbon dioxide and toxic smoke, offsetting green energy greenhouse gas emission reductions.

NIMBYism also makes it hard to erect unsightly gray infrastructure such as levees around residential neighborhoods. In some cases this can be resolved with "natural capital" solutions such as horizontal levels and restored marshlands without detracting from an area's natural appeal. But in other circumstances, particularly around critical power plants and wastewater facilities, those kinds of natural capital solutions may not be practical or feasible.

Gray solutions such as dams and reservoirs that control flooding can have negative consequences for wildlife, a legitimate environmental concern. The layering of multiple permitting processes on a given project gives considerable bargaining leverage to groups that use environmental objections to keep unsightly projects out of their neighborhoods and to nonprofit wildlife organizations that are sincerely trying to minimize harm to birds and desert turtles. Delays that happen while these issues are worked out can leave areas vulnerable to flooding or wildfires at risk for years as permitting processes work their way through the maze of state and federal agency review. This is what happened in the case of San Francisquito Creek, as discussed in chapter 3. The bottom line is that local communities will have many nonpartisan issues to resolve related to building resilience.

5. Approaches to both decarbonization and resilience have to become more politically strategic to be successful.

Taking full measure of all the steps that have to be taken by individuals and communities to decarbonize the economy and build resilience will entail enormous expense and considerable disruption. Businesses that electrify their trucks and manufacturing processes may place themselves at competitive disadvantage to out-of-state firms that operate with fewer constraints. Consumers may face myriad expenses in order to make their homes more resilient to higher temperatures, wildfire smoke, and floods while at the same time being forced to replace their gas stoves, switch out their heating systems, and buy new electric cars. Local governments must do the same with their buildings, cars, and trucks while finding the resources to pay for green and gray infrastructure that will adequately protect their residents and businesses. All of this is a big ask with lots of potential political risk for those who must do the asking.

The context for all of this matters. More partisan polarization will make these essential tasks harder, but deteriorating weather conditions should make the urgency of acting on climate policy more apparent. The current modeling that government agencies get from consulting firms and academics usefully project out possible technical options and their costs. But the people who make these models do not factor into their calculations the political constraints that would guide public officials toward the best feasible path forward. This responsibility is left to legislators, lobbyists, and other political actors who often have their own agendas.

There needs to be a lot of learning by doing with respect to the design and implementation of climate change mitigation and adaptation technology. Ideally, this means monitoring and truthfully reporting the results of policy designs and learning from mistakes. But too often agencies and elected officials "can't handle the truth" because they are fearful of being blamed for failure and losing an election or being removed from agency leadership. This can lead to a chain of truthiness problems and greenwashing that prevents progress and propagates faith in flawed policies, delaying efforts to find new solutions. There needs to be a stronger commitment to sharing data and full transparency about policies and their impacts.

6. Making policy in emergency mode fixes some resilience prob-
    lems but creates others.

One example of effective action is that of drought emergencies. Under emergency decrees, governors can waive requirements that hold up solutions and impose guidelines on water usage that voluntary efforts could never achieve. But emergency powers are contingent for good reasons. They can be problematic in various ways. Emergency actions can be extended to matters that are only indirectly or even peripherally related to the drought crisis. They are also temporary, and when the emergency ends, people and organization can revert to their old habits.

Crises that go on for lengthy periods of time disrupt the balance of power between the executive and other government branches. When there are wars, patriotism kicks in, and most people are committed to the cause of victory, not looking to exploit the system. But when emergency powers are imposed on a divided electorate over extensive periods of time, this can lead to resistance and civil unrest. The length of the COVID-19 pandemic crisis has hardened political lines and fed the collective paranoia of those who believe that pandemics are ruses for elites who want more government control over people's lives. Such paranoia subverts government legitimacy and can lead to the kind of events we witnessed on January 6, 2021.

FINAL THOUGHTS

The peculiar design of US government favors the status quo and forces the country to build political consensus either across party lines or within a large majority coalition in order to undertake transformative policy. Crises can open windows of opportunity that enable major policy shifts if the conditions are right. But when people sense that transformational change is necessary and the system is incapable of providing it, resistance can morph into changing the system in order to enable the policy. Both climate change policy and the prospect of ongoing pandemics could challenge democratic systems in a fundamental way.

The American system, with its multiple checks and balances, favors the status quo over the kind of transformative change that climate policies require. Stable winning strategies are based on building broad coalitions across party, regional, and socioeconomic divides. A successful climate strategy is one that balances the urgency of action with the time and effort needed to build trust and confidence. Politically, we might have made more progress if there had not been such a long lag between what was happening in the atmosphere and what was happening on the ground.

But there are good reasons to believe that working toward resilience now could help build trust needed to take on the harder tasks of decarbonization later. Weather-related crises might indeed induce the compromise and build the consensus needed to meet the climate change challenge. In this regard, extreme weather might ultimately lead to better climate policies.

ACKNOWLEDGMENTS

There are many people to thank for their influence and assistance in writing this book. Much of the background material about the western region draws on my experience of teaching an American West class at Stanford University since 2013. So, I owe my colleagues David Kennedy, David Freyberg, Alex Nemerov, and Shelley Fisher Fishkin for what each of them has taught me about the distinctive history, culture, and environment of the West. Next, I thank Surabhi Balachander for her skillful help with cleaning up and preparing the manuscript for publication. My now-deceased colleague Iris Hui aided me in designing the regional survey and coauthoring many of the journal papers that ultimately led me in the direction of writing this book. She is missed very much by all of us at the Bill Lane Center for the American West. I received encouragement and feedback from my colleagues Kylie Gordon, Kate Gibson, Celina Maya Scott-Buechler, Emily Zhang, Denis Lacorne, and Felicity Barringer, all of whom read and reacted to early drafts. Last but not least, I want to thank the Foster clan for attending the lectures in 2019 and sharing their perspectives over the years on Texas and Oklahoma: Tom and Amy Foster, Barbara Measelle and Mike Foster, and my wife, Jody Foster, who also came up with the title for the book.

APPENDIX

BILL LANE CENTER SURVEYS INFORMATION

BILL LANE CENTER WESTERN STATES REGIONAL SURVEY, AUGUST–SEPTEMBER 2019, WITH ENLARGED CALIFORNIA SUBSAMPLE

YouGov interviewed 3,195 respondents who were then matched down to a sample of 3,000 to produce the final data set. Within the regional sample, we had a subsample of 1,042 respondents in California. The respondents were matched to a sampling frame on gender, age, race, and education. The frame was constructed by stratified sampling from the western states of the 2016 American Community Survey (ACS) one-year sample, with selection within strata by weighted sampling with replacements (using the person weights on the public-use file). The 2019 wildfire questions were part of the regional poll. The regional poll also contained questions about different facets of living in the state, including length of residence and what respondents like or dislike about their state.

BILL LANE CENTER CALIFORNIA WILDFIRE SURVEY, 2020

YouGov interviewed 1,646 respondents in California who were then matched down to a sample of 1,500 to produce the final data set. The respondents were matched to a sampling frame on gender, age, race, and education. The frame was constructed by stratified sampling from the full 2018 ACS one-year sample, with selection within strata by weighted sampling with replacements (using the person weights on the public-use file). The matched cases were weighted to the sampling frame using propensity scores. The matched cases and the frame were combined, and a logistic regression was estimated for inclusion in the frame. The propensity score function included age, gender, race/ethnicity, years of education, and region. The propensity scores were grouped into deciles of the estimated propensity score in the frame and poststratified according to these deciles. The weights were then poststratified on 2016 presidential vote choice, and a four-way stratification of gender, age (four categories), race (four categories), and education (four categories) was used to produce the final weight.

HOOVER INSTITUTION–BILL LANE CENTER WATER POLICY SURVEY, 2015

YouGov interviewed 1,664 respondents who were then matched down to a sample of 1,500 to produce the final data set. The respondents were matched to a sampling frame on gender, age, race, education, and geographic location in California. The frame was constructed by stratified sampling from the full 2013 ACS sample, with selection within strata by weighted sampling with replacements (using the person weights on the public-use file). The matched cases were weighted to the sampling frame using propensity scores. The matched cases and the frame were combined, and a logistic regression was estimated for inclusion in the frame. The propensity score function included age, gender, race/ethnicity, years of education, and geographical region. The propensity scores were grouped into deciles of the estimated propensity score in the frame and poststratified according to these deciles.

BILL LANE CENTER SEA LEVEL RISE, 2019

YouGov interviewed 1,718 respondents who were then matched down to a sample of 1,500 to produce the final data set. The respondents were matched to a sampling frame on gender, age, race, and education. The frame was constructed by stratified sampling from the California portion of the 2016 ACS one-year sample, with selection within strata by weighted sampling with replacements (using the person weights on the public-use file). The matched cases were weighted to the sampling frame using propensity scores. The matched cases and the frame were combined, and a logistic regression was estimated for inclusion in the frame. The propensity score function included age, gender, race/ethnicity, years of education, and region. The propensity scores were grouped into deciles of the estimated propensity score in the frame and poststratified according to these deciles. The weights were then poststratified on 2016 presidential vote choice, and a four-way stratification of gender, age (four categories), race (four categories), and education (four categories) was used to produce the final weight.

NOTES

INTRODUCTION

1. Richard H. Dillon, "Stephen Long's Great American Desert," *Proceedings of the American Philosophical Society* 111, no. 2 (1967): 93–108.

2. For a review of the literature on the political strains on the US system and how this has impacted policy making, see Morris P. Fiorina, *Unstable Majorities: Polarization, Party Sorting, and Political Stalemate* (Hoover Institution Press, 2017); see also Frances E. Lee, *Insecure Majorities: Congress and the Perpetual Campaign* (University of Chicago Press, 2016).

3. Bruce Cain and Francis Fukuyama, "Our Peculiar Reform Challenge," *American Interest* (October 2015), https://www.the-american-interest.com/2015/10/15/our-peculiar-reform-challenge/.

4. Riley E. Dunlap, Aaron M. McCright, and Jerrod H. Yarosh, "The Political Divide on Climate Change: Partisan Polarization Widens in the US," *Environment: Science and Policy for Sustainable Development* 58, no. 5 (2016): 4–23.

5. The expansion of these democratic opportunities is often based on naive assumptions about human behavior. See Bruce E. Cain, *Democracy More or Less* (Cambridge University Press, 2015).

6. Francis Fukuyama, "American Political Decay or Renewal: The Meaning of the 2016 Election," *Foreign Affairs* 95 (2016): 58.

7. John W. Kingdon and Eric Stano, *Agendas, Alternatives, and Public Policies*, Vol. 45 (Little, Brown, 1984).

8. Gerald D. Nash, *The American West Transformed: The Impact of the Second World War* (University of Nebraska Press, 1990), and, more recently, Mark Brilliant and David M. Kennedy, eds., *World War II and the West It Wrought* (Stanford University Press, 2020).

9. Ellen Hanak et al., *Managing California's Water: From Conflict to Resolution*, Public Policy Institute of California, 2011, https://www.ppic.org/wp-content/uploads/content/pubs/report/R_211EHR.pdf.

10. Richard B. Stewart, "Regulation, Innovation, and Administrative Law: A Conceptual Framework," *California Law Review* 69 (1981): 1256–377.

11. Bruce E. Cain, Russell J. Dalton, and Susan E. Scarrow, eds., *Democracy Transformed? Expanding Political Opportunities in Advanced Industrial Democracies* (Oxford University Press on Demand, 2006).

12. Jessica M. Ayers and Saleemul Huq, "The Value of Linking Mitigation and Adaptation: A Case Study of Bangladesh," *Environmental Management* 43, no. 5 (2009): 753–64.

13. For a review of the literature on partisanship as it affects resilience and an illustration with data on sea level rise in California, see Bruce E. Cain, Elisabeth R. Gerber, and Iris Hui, "Getting Bipartisan Support for Sea Level Rise Adaptation Policies," *Ocean & Coastal Management* 197 (2020), https://doi.org/10.1016/j.ocecoaman.2020.105298.

14. Jonas Colliander, "'This Is Fake News': Investigating the Role of Conformity to Other Users' Views When Commenting on and Spreading Disinformation in Social Media," *Computers in Human Behavior* 97 (2019): 202–15.

15. NIMBY is an acronym for the phrase "not in my back yard" that refers to neighborhood objections to siting commercial or infrastructure projects near their homes.

16. Jiazhe Sun and Kaizhong Yang, "The Wicked Problem of Climate Change: A New Approach Based on Social Mess and Fragmentation," *Sustainability* 8, no. 12 (2016), https://doi.org/10.3390/su8121312.

17. Sheldon Ungar, "Knowledge, Ignorance and the Popular Culture: Climate Change versus the Ozone Hole," *Public Understanding of Science* 9, no. 3 (2000): 297–312.

18. Frederike Albrecht and Charles F. Parker, "Healing the Ozone Layer: The Montreal Protocol and the Lessons and Limits of a Global Governance Success Story," in *Great Policy Successes*, ed. Paul 't Hart, 304–22 (Oxford University Press, 2019).

19. Michael Burger, Jessica Wentz, and Radley Horton, "The Law and Science of Climate Change Attribution," *Columbia Journal of Environmental Law* 45 (2020), https://doi.org/10.7916/cjel.v45i1.4730.

20. Patrick J. Egan and Megan Mullin, "Turning Personal Experience into Political Attitudes: The Effect of Local Weather on Americans' Perceptions about Global Warming," *Journal of Politics* 74, no. 3 (2012): 796–809. For their review of the literature on how short-term weather affects climate change attitudes, see Patrick J. Egan and Megan Mullin, "Climate Change: US Public Opinion," *Annual Review of Political Science* 20 (2017): 209–27.

CHAPTER 1. DISTINCTIVE WESTERN FEATURES

1. "View of the World from 9th Avenue," Cover illustration, *New Yorker*, March 29, 1976.

2. Kevin Krajiick, "The 100th Meridian, Where the Great Plains Begin, May Be Shifting," Columbia Climate School, April 11, 2018, https://news.climate.columbia.edu/2018/04/11/the-100th-meridian-where-the-great-plains-used-to-begin-now-moving-east.

3. J. W. Powell, *Report on the Arid Region of the United States* (US Geological Survey, 1878).

4. Harrison C. Dunning, "Dam Fights and Water Policy in California: 1969–1989," Innovation in Western Water Law and Management, Summer Conference, June 5–7, 1991, https://scholar.law.colorado.edu/innovation-in-western-water-law-and-management/19/?utm_source=scholar.law.colorado.edu%2Finnovation-in-western-water-law-and-management%2F19&utm_medium=PDF&utm_campaign=PDFCoverPages.

5. Ifran Alvi, "Case Study: Oroville Dam (California, 2017)," Association of State Dam Safety Officials, 2017, https://damfailures.org/case-study/oroville-dam-california-2017/.

6. Josh Clemons, "Interstate Water Disputes: A Road Map for States," *Southeastern Environmental Law Journal*12 (2003): 115.

7. *Oklahoma v Texas*, 260 US 606 (1923) and *Tarrant Regional Water District v Herrmann* 569 YS 614 (2013).

8. See word cloud graphics in Bruce E. Cain, "Climate Politics in the West: Is Red the New Green?," YouTube, October 22, 2019, 2019, https://www.youtube.com/watch?v=NepluMPecB4.

9. "Oil and Gas Producing States," Earthworks, 2021, https://earthworks.org/issues/oil_and_gas_producing_states/.

10. "Which States Produce the Most Wind Energy," Inspire Clean Energy, October 2020, https://www.inspirecleanenergy.com/blog/clean-energy-101/which-states-produce-the-most-wind-energy.

11. Robert T. Anderson, "Indian Water Rights and the Federal Trust Responsibility," *Natural Resources Journal* 46 (2006): 399.

12. "Health and Healthcare in Frontier Areas," Rural Health Information Hub, November 2, 2022, https://www.ruralhealthinfo.org/topics/frontier.

CHAPTER 2. CLIMATE CREEP AND
DROUGHT POLICY CYCLES

1. "Historical Data and Conditions," National Integrated Drought Information System, n.d., https://www.drought.gov/historical-information?state=california&dataset=0&selectedDateUSDM=20211207.

2. Carey McWilliams, *California: The Great Exception* (University of California Press, 1949).

3. Edward R. Cook, Richard Seager, Mark A. Cane, and David W. Stahle, "North American Drought: Reconstructions, Causes, and Consequences," *Earth-Science Reviews* 81, nos. 1–2 (2007): 93–134.

4. Susan E. Swanberg, "'The Way of the Rain': Towards a Conceptual Framework for the Retrospective Examination of Historical American and Australian 'Rain Follows the Plow/Plough' Messages," *Human Ecology Review* 25, no. 2 (2019): 67–95.

5. Stephen N. Bretsen and Peter J. Hill, "Irrigation Institutions in the American West," *UCLA Journal of Environmental Law and Policy* 25 (2006): 283.

6. Cheryl A. Dieter, Molly A. Maupin, Rodney R. Caldwell, Melissa A. Harris, Tamara I. Ivahnenko, John K. Lovelace, Nancy L. Barber, and Kristin S. Linsey, *Water Availability and Use Science Program: Estimated Use of Water in the United States in 2015* (US Geological Survey, 2018).

7. "Economic Impact of Agriculture," University of Arkansas Division of Agriculture, n.d., https://economic-impact-of-ag.uada.edu/.

8. John A. Ferejohn, *Pork Barrel Politics: Rivers and Harbors Legislation, 1947–1968* (Stanford University Press, 1974).

9. Chris Edwards and Peter J. Hill, "Cutting the Bureau of Reclamation and Reforming Water Markets," Downsizing the Federal Government, February 1, 2012, https://www.downsizinggovernment.org/interior/cutting-bureau-reclamation.

10. Jedidiah Brewer et al., "Water Markets in the West: Prices, Trading, and Contractual Forms," National Bureau of Economic Research, Working Paper 13002, March 2007.

11. "History," San Diego County Water Authority, 2022, https://www.sdcwa.org/about-us/history/.

12. Ian James, "Desalination Is the Costliest Solution, Study Finds," *Desert Sun*, October 14, 2016.

13. Lisa Krieger, "'A State of Drought': Coachella Valley Grapples with Shrinking Water Supply, *Santa Cruz Sentinel*, October 3, 2015.

14. H. Stuart Burness and James P. Quirk, "Appropriative Water Rights and the Efficient Allocation of Resources," *American Economic Review* 69, no. 1 (1979): 25–37.

15. Alex Dobuzinskis, "California Curtails Some Longstanding Water Rights over Drought," Reuters, June 12, 2015.

16. "Hoover Institution Golden State Poll," Hoover Institution, September 2015, https://www.hoover.org/hoover-institution-golden-state-poll.

17. Rafe Petersen and Daniel R. Golub, "Obama Administration Issue Final Rule on Waters of the United States," Holland & Knight, June 9, 2015, https://www.hklaw.com/en/insights/publications/2015/06/obama-administration-issues-final-rule-on-waters-0.

18. Judith S. Hendry, J. Delicath Depoe, and M. Elsenbeer, "Decide, Announce, Defend: Turning the NEPA Process into an Advocacy Tool Rather Than a

Decision-Making Tool," in *Communication and public participation in environmental decision-making*, ed. Stephen P. Depoe, John W. Delicath, and Marie-France Aepli Elsenbeer, 99–112 (SUNY Press, 2004).

19. Bruce E. Cain, *Democracy More or Less* (Cambridge University Press, 2015), chap. 3.

20. Cain, *Democracy More or Less*, chap. 2.

21. Kathleen Bawn, "Political Control versus Expertise: Congressional Choices about Administrative Procedures," *American Political Science Review* 89, no. 1 (1995): 62–73.

22. S. Abraham Ravid, Kose John, Balbinder Singh Gill, and Jongmoo Jay Choi, "Polls, Politics and Disaster Relief: Evidence from Federal SBA Loan Programs," SSRN, March 4, 2021), https://ssrn.com/abstract=3797341.

23. National Drought Policy Commission, *Preparing for Drought in the 21st Century* (US Department of Agriculture, May 2000).

24. Andrew Healy and Neil Malhotra, "Myopic Voters and Natural Disaster Policy," *American Political Science Review* 103, no. 3 (2009): 387–406.

25. "Western Governors' Drought Forum: Special Report," Western Governors' Association, June 2015, https://westgov.org/images/files/FINAL_Drought_Forum _June_2015.pdf.

26. Compiled from data from the California Secretary of State.

27. Heather Cooley, "Urban and Agricultural Water Use in California, 1960–2015," Pacific Institute, May 2020), https://pacinst.org/publication/urban-agricultural -water-use/.

28. "2015 Water Issues Public Opinion Poll," San Diego County Water Authority, April 2015, https://www.sdcwa.org/sites/default/files/files/news-center/2015-public -opinion-poll.pdf.

29. Iris Hui and Bruce E. Cain, "Overcoming Psychological Resistance toward Using Recycled Water in California," *Water and Environment Journal* 32, no. 1 (2018): 17–25.

30. Cassidy Craford and Hannah Safford, "When Do Water Bonds Pass?," California WaterBlog, April 14, 2019, https://californiawaterblog.com/2019/04/14/when -do-water-bonds-pass-evidence-from-past-elections/.

31. Sarah Michaels, Nancy P. Goucher, and Dan McCarthy, "Policy Windows, Policy Change, and Organizational Learning: Watersheds in the Evolution of Watershed Management," *Environmental Management* 38, no. 6 (2006): 983–92.

32. Kimberly J. Quesnel and Newsha K. Ajami, "Changes in Water Consumption Linked to Heavy News Media Coverage of Extreme Climatic Events," *Science Advances* 3, no. 10 (2017): e1700784.

33. Sam Sanders, "In California, Technology Makes Droughtshaming Easier Than Ever," National Public Radio, May 25, 2015.

34. Kimberly J. Quesnel, Saahil Agrawal, and Newsha K. Ajami, "Diverse Para-digms of Residential Development Inform Water Use and Drought-Related

Conservation Behavior," *Environmental Research Letters* 15, no. 12 (2020), https://iopscience.iop.org/article/10.1088/1748-9326/abb7ae/pdf.

35. Caitrin Chappelle, Ellen Hanak, and Thomas Harter, "Groundwater in California," Public Policy Institute of California, May 2017, https://www.ppic.org/wp-content/uploads/JTF_GroundwaterJTF.pdf.

CHAPTER 3. WILDFIRE TERROR AND POLICY SPARKS

1. Tim Arango, Johnny Diaz, and Carly Stern, "Three Killed in Fresh Wildfires in Northern California," *New York Times*, September 29, 2020, https://www.nytimes.com/2020/09/28/us/california-glass-zogg-fires.html.

2. Arango, Diaz, and Stern.

3. California Air Resources Board, "Greenhouse Gas Emissions of Contemporary Wildfire, Prescribed Fire and Forest Management Activities," December 2020.

4. "Benefits of Fire," CalFire, n.d., https://www.fire.ca.gov/media/5425/benifitsoffire.pdf.

5. "Wildland Fire Exposure," Stanford University, n.d., https://climatehealth.sites.stanford.edu/wildland-fire-exposure.

6. Volker C. Radeloff et al., "Rapid Growth of the US Wildland-Urban Interface Raises Wildfire Risk," *Proceedings of the National Academy of Sciences* 115, no. 13 (2018): 3314–19.

7. Radeloff et al.

8. Rebecca K. Miller, Christopher B. Field, and Katharine J. Mach, "Barriers and Enablers for Prescribed Burns for Wildfire Management in California," *Nature Sustainability* 3, no. 2 (2020): 101–9.

9. Rebecca K. Miller, "Prescribed Burns in California: A Historical Case Study of the Integration of Scientific Research and Policy," *Fire* 3, no. 3 (2020): 44.

10. Julie Cart, "The West Is Burning, So California Struggles to Find Help Fighting Its Fires," CalMatters, August 28, 2020, https://calmatters.org/environment/california-wildfires/2020/08/california-wildfires-help-neighbors/.

11. "California Is Not Prepared to Protect Its Most Vulnerable Residents from Natural Disasters," California State Auditor, December 2019, https://www.auditor.ca.gov/pdfs/reports/2019-103.pdf.

12. James Rufus, "Insurer Merced Went Belly Up after Camp Fire," *Los Angeles Times*, December 4, 2018.

13. Nicole Friedman, "High Cost of Wildfire Insurance Hurts California Home Sales," *Wall Street Journal*, January 5, 2020.

14. "NWCG Report on Wildland Firefighter Fatalities in the United States: 2007–2016," National Wildfire Coordinating Group, December 2017, https://www.nwcg.gov/sites/default/files/publications/pms841.pdf.

15. Gabrielle Levy, "Wildfires Are Getting Worse, and More Costly, Every Year," *U.S. News & World Report*, August 1, 2018.

16. Headwaters Economics, "Montana Wildfire Cost Study: Technical Report," August 8, 2008, https://headwaterseconomics.org/wp-content/uploads /HeadwatersEconomics_FireCostStudy_TechnicalReport.pdf.

17. Bella Isaacs-Thomas, "California's Catastrophic Wildfires in Three Charts," PBS, September 14, 2020, https://www.pbs.org/newshour/science/californias -catastrophic-wildfires-in-3-charts/.

18. For instance, experiences with flooding seem to affect attitudes about clime change. See Christina Demski, Stuart Capstick, Nick Pidgeon, Robert Gennaro Sposato, and Alexa Spence, "Experience of Extreme Weather Affects Climate Change Mitigation and Adaptation Responses," *Climatic Change* 140, no. 2 (2017): 149–64. But high temperatures and precipitation levels did not affect attitudes about climate change. See Jennifer R. Fownes and Shorna B. Allred, "Testing the Influence of Recent Weather on Perceptions of Personal Experience with Climate Change and Extreme Weather in New York State," *Weather, Climate, and Society* 11, no. 1 (2019): 143–57. The evidence in the literature seems to suggest that direct experience with natural disasters or perceived harm impacts climate attitudes the most. See Michalis Diakakis, Michalis Skordoulis, and Eleni Savvidou, "The Relationships between Public Risk Perceptions of Climate Change, Environmental Sensitivity and Experience of Extreme Weather-Related Disasters: Evidence from Greece," *Water* 13, no. 20 (2021), https://doi.org/10.3390/w13202842; and Chad Zanocco, Hilary Boudet, Roberta Nilson, and June Flora, "Personal Harm and Support for Climate Change Mitigation Policies: Evidence from 10 US Communities Impacted by Extreme Weather," *Global Environmental Change* 59 (2019), https:// doi.org/10.1016/j.gloenvcha.2019.101984.

19. Iris Hui, Angela Zhao, Bruce E. Cain, and Anne M. Driscoll. "Baptism by Wildfire? Wildfire Experiences and Public Support for Wildfire Adaptation Policies," *American Politics Research* 50, no. 1 (2022):108–16.

CHAPTER 4. RELUCTANT RESILIENCE

1. Anne Mulkern, "In California Rising Seas Pose a Bigger Economic Threat than Wildfires, Quakes," *Scientific American*, March 14, 2019, https://www .scientificamerican.com/article/in-california-rising-seas-pose-a-bigger-economic -threat-than-wildfires-quakes/.

2. Brad Plumer and Nadja Popovich, "Yes There Has Been Progress on Climate, and No, It's Not Nearly Enough," *New York Times*, October 2021.

3. T. Moon, D. A. Sutherland, D. Carroll, D. Felikson, L. Kehrl, and F. Straneo, "Subsurface Iceberg Melt Key to Greenland Fjord Freshwater Budget," *Nature Geoscience* 11, no. 1 (2018): 49–54.

4. Ocean Studies Board and National Research Council, *Sea-Level Rise for the Coasts of California, Oregon, and Washington: Past, Present, and Future* (National Academies Press, 2012).

5. "Preparing for Rising Seas: How the State Can Help Support Local Coastal Adaptation Efforts," California Legislative Analyst's Office, December 2019, https://lao.ca.gov/reports/2019/4121/coastal-adaptation-121019.pdf.

6. Matthew Heberger, Heather Cooley, Eli Moore, and Pablo Herrera, "The Impacts of Sea Level Rise on the San Francisco Bay," Pacific Institute, July 2012, https://pacinst.org/wp-content/uploads/2018/08/sea_level_rise_sf_bay_cec3.pdf.

7. Rosanna Xia, "The California Coast Is Disappearing under the Rising Sea," *Los Angeles Times*, July 2019. See also "What Threat Does Sea Level Rise Pose to California?," California Legislative Analyst's Office, August 10, 2020, https://lao.ca.gov/Publications/Report/4261.

8. "Preparing for Rising Seas."

9. Mary Sprague, Kate F. Wilson, and Bruce E. Cain, "Reducing Local Capacity Bias in Government Grantsmanship," *American Review of Public Administration* 49, no. 2 (2019): 174–88.

10. The author had multiple interviews with San Francisquito Creek Joint Powers Authority executive director Len Materman from 2014 to 2016.

11. Gennady Sheyner, "Water Board Deals a Blow to Flood Control Effort," Palo Alto Online, March 4, 2014, https://paloaltoonline.com/news/2014/03/04/water-board-deals-a-blow-to-flood-control-effort.

12. Sheyner.

13. Henry E. Brady, Sidney Verba, and Kay Lehman Schlozman, "Beyond SES: A Resource Model of Political Participation," *American Political Science Review* 89, no. 2 (1995): 271–94.

14. Kent Barnett, "Why Bias Challenges to Administrative Adjudication Should Succeed," *Missouri Law Review* 81, no. 4 (2016): 1023–44.

15. Jennifer L. Hernandez, David Friedman, and Stephanie DeHerrera , *In the Name of the Environment: Litigation Abuse Under CEQA*, Holland & Knight, 2015, https://issuu.com/hollandknight/docs/ceqa_litigation_abuseissuu?e=16627326/14197714.

16. Hernandez, Friedman, and DeHerrera.

17. Mark Brilliant and David M. Kennedy, eds., *World War II and the West It Wrought* (Stanford University Press, 2020).

18. N. Ray Gilmore and Gladys W. Gilmore, "The Bracero in California," *Pacific Historical Review* (1963): 265–82.

19. For more on the meaning of geographic sorting, see Wendy K. Tam Cho, James G. Gimpel, and Iris S. Hui, "Voter Migration and the Geographic Sorting of the American Electorate," *Annals of the Association of American Geographers* 103, no. 4 (2013): 856–70.

20. Bruce E. Cain, Elisabeth R. Gerber, and Iris Hui, "Getting Bipartisan Support for Sea Level Rise Adaptation Policies," *Ocean & Coastal Management* 197 (2020), https://doi.org/10.1016/j.ocecoaman.2020.105298.

21. Cain et al.

22. Cain et al.

23. Priscilla DeGregory, "NYC's Partially Halted Anti-Flood Plan Can Go Forward," *New York Post*, November 30, 2021.

24. Cain et al., "Getting Bipartisan Support for Sea Level Rise Adaptation Policies."

25. Iris Hui, Gemma Smith, and Caroline Kimmel, "Think Globally, Act Locally: Adoption of Climate Action Plans in California," *Climatic Change* 155, no. 4 (2019): 489–509.

26. Rebecca Nelson, "A City Rose on the Marshes, Will the Bay Take It Back," *& the West*, December 19, 2018, https://andthewest.stanford.edu/2018/a-city-rose-on-the-marshes-will-the-bay-take-it-back.

27. Mark Lubell and Matthew Robbins, "Adapting to Sea-Level Rise: Centralization or Decentralization in Polycentric Governance Systems?," *Policy Studies Journal* 49, no. 2 (2021): 562–90.

CHAPTER 5. THE PATHS OF WATER
AND ENERGY GOVERNANCE

1. Sergio Fabbrini, *Compound Democracies: Why the United States and Europe Are Becoming Similar* (Oxford University Press, 2010).

2. E. Norman Veasey, "What Would Madison Think: The Irony of the Twists and Turns of Federalism," *Delaware Journal of Corporate Law* 34, no. 1 (2009): 35–56. With regard to energy policy, see Scott Jacobson, "Dual Sovereignty Is Out, Time for Concurrent Jurisdiction to Shine," *William & Mary Environmental Law and Policy Review* 42, no. (2018): 627–45.

3. Economists have grappled for many years with the issue of how to regulate natural monopolies. See, for instance, Hayne E. Leland, "Regulation of Natural Monopolies and the Fair Rate of Return," *Bell Journal of Economics and Management Science* (1974): 3–15.

4. Eric J. OShaughnessy, Jenny S. Heeter, Julien Gattaciecca, Jennifer Sauer, Kelly Trumbull, and Emily I. Chen, "Community Choice Aggregation: Challenges, Opportunities, and Impacts on Renewable Energy Markets," National Renewable Energy Lab, February 2019, https://www.nrel.gov/docs/fy19osti/72195.pdf.

5. "Water Special Districts: A Look at Governance and Public Participation," California Legislative Analyst Office, March 2002, https://lao.ca.gov/2002/water_districts/special_water_districts.html.

6. Vanessa Casado-Pérez, Bruce E. Cain, Iris Hui, Coral Abbott, Kaley Dodson, and Shane Lebow, "All Over the Map: The Diversity of Western Water Plans," *California Journal of Politics and Policy* 7, no. 2 (2015), http://dx.doi.org/10.5070/P2cjpp7225762.

7. See Iris Hui, Nicola Ulibarri, and Bruce E. Cain, "Patterns of Participation and Representation in a Regional Water Collaboration," *Policy Studies Journal* 48, no. 3 (2020): 754–81; and Bruce E. Cain, Elisabeth R. Gerber, and Iris Hui, "The

Challenge of Externally Generated Collaborative Governance: California's Attempt at Regional Water Management," *American Review of Public Administration* 50, nos. 4–5 (2020): 428–37.

8. Parth Vaishnav, Nathaniel Horner, and Inês L. Azevedo, "Was It Worthwhile? Where Have the Benefits of Rooftop Solar Photovoltaic Generation Exceeded the Cost?," *Environmental Research Letters* 12, no. 9 (2017): 094015.

9. Christopher Butler, "Electric Vehicles Prices Finally in Reach of Millennial, Gen Z Car Buyers," CNBC, October 21, 2019, https://www.cnbc.com/2019/10/20 /electric-car-prices-finally-in-reach-of-millennial-gen-z-buyers.html.

10. "Consumers Purchasing an Electric Vehicle Are Younger and More Affluent Than Those Buying a Hybrid," Experian, April 22, 2014, https://www.experian.com /blogs/news/2014/04/22/consumers-purchasing-an-electric-vehicle-are-younger -and-more-affluent-than-those-buying-a-hybrid/.

11. "NREL Report Sheds Light on Community Choice Aggregation in the United States," National Renewable Energy Laboratory, July 8, 2019, https://www.nrel.gov /news/program/2019/report-sheds-light-on-community-choice-aggregation-in-the -united-states.html.

12. Morris Fiorina, ed., *Who Governs? Emergency Powers in the Time of COVID* (Hoover Institution Press, 2023).

INDEX